SRSK5T09

TRACE ORGANIC CHEMICALS IN BIOSOLIDS-AMENDED SOILS

STATE-OF-THE-SCIENCE REVIEW

by:

Christopher P. Higgins
Jonathan O. Sharp
Jennifer G. Sepulvado
BobbiJo Littrell
Colorado School of Mines

George O'Connor
University of Florida

Erin Snyder
Total Environmental Solutions, Inc.

Drew McAvoy
The Procter & Gamble Company

2010

WERF

The Water Environment Research Foundation, a not-for-profit organization, funds and manages water quality research for its subscribers through a diverse public-private partnership between municipal utilities, corporations, academia, industry, and the federal government. WERF subscribers include municipal and regional water and wastewater utilities, industrial corporations, environmental engineering firms, and others that share a commitment to cost-effective water quality solutions. WERF is dedicated to advancing science and technology addressing water quality issues as they impact water resources, the atmosphere, the lands, and quality of life.

For more information, contact:
Water Environment Research Foundation
635 Slaters Lane, Suite G-110
Alexandria, VA 22314-1177
Tel: (571) 384-2100
Fax: (703) 299-0742
www.werf.org
werf@werf.org

This report was co-published by the following organization.

IWA Publishing
Alliance House, 12 Caxton Street
London SW1H 0QS, United Kingdom
Tel: +44 (0) 20 7654 5500
Fax: +44 (0) 20 7654 5555
www.iwapublishing.com
publications@iwap.co.uk

Library of Congress Catalog Card Number: 2010921255
IWAP ISBN: 978-1-84339-393-1/1-84339-393-x

Colorado School of Mines

ACKNOWLEDGMENTS

The research teams thanks the members of the Project Subcommittee (PSC) for their constructive critique and feedback. The individuals volunteering their time in this advisory role are listed below. The team also thanks the external peer reviewers that agreed to review a provide comments on this report: Steve Beaulieu, Thomas Borch, Edward T. Furlong, Roman P. Lanno, Linda S. Lee, David Quanrud, Kate M. Scow, and Tracy Yager. Thanks also go to Joseph Wiley, a Colorado School of Mines graduate student who was diligent in his pursuit of data.

Report Preparation

Principal Investigators:
Christopher P. Higgins, Ph.D.
Jonathan O. Sharp, Ph.D.
Colorado School of Mines

Project Team:
James Gray, Ph.D.
United States Geological Survey

Lakhwinder Hundal, Ph.D.
Metropolitan Water Reclamation District of Greater Chicago

Drew McAvoy, Ph.D.
The Procter & Gamble Company

George O'Connor, Ph.D.
University of Florida

Jennifer G. Sepulvado, M.S.
BobbiJo Littrell
Colorado School of Mines

Erin Snyder, Ph.D.
Total Environmental Solutions, Inc.

Edward Topp, Ph.D.
Agriculture and Agri-Food Canada

Project Subcommittee

Thomas Granato
Metropolitan Water Reclamation District of Greater Chicago

Greg Kester
California Association of Sanitation Agencies

Mark Ramirez
District of Columbia, Water and Sewer Authority

Rick Stevens
U.S. Environmental Protection Agency

David S. Taylor
Madison Metropolitan Sewerage District

Water Environment Research Foundation Staff

Director of Research:	Daniel M. Woltering, Ph.D.
Program Manager:	Alan Hais, P.E.

ABSTRACT AND BENEFITS

Abstract:

The presence of trace organic chemicals (TOrCs) in municipal biosolids in the U.S. has received considerable attention by the public and scientific community over the last several years. Of particular concern is whether the presence of TOrCs in biosolids results in significant risks to public health and the environment following land application. The U.S. Environmental Protection Agency evaluated the risks associated with dioxins present in biosolids-amended soils, but assessments of TOrCs of emerging concern have not been similarly assessed, owing in part to limited data needed for the risk assessment. In this study, an evaluation was made to determine the TOrCs of greatest concern in the terrestrial environment, categorized as high priority and low priority TOrCs. The assessment was based on occurrence data and readily available information on basic properties such as bioaccumulation and toxicity. An evaluation of quantitative risk assessments was also conducted to identify the most important parameters for conducting ecological risk assessments and the techniques currently available for obtaining the parameter values. A minimum data set for risk modeling was identified. A comprehensive literature review of the identified TOrCs of greatest concern identified relevant data on fate, transport, biotransfer from soil to plants and animals, and toxicity in the terrestrial environment. Based on the review, data gaps were identified for the parameters most important for conducting terrestrial risk assessments.

Benefits:

- Identifies the TOrCs of potential greatest concern for the land application of biosolids and prioritized them based on occurrence data and readily available data on bioaccumulation and toxicity.
- Provides a comprehensive compilation of biosolids occurrence data for the targeted TOrCs.
- Provides an examination of risk assessment methodology used in the United States and Europe and identifies the minimum data set needed for ecological and human risk assessment modeling.
- Provides a detailed overview of what is currently known about the physical, chemical, and biological processes affecting TOrC fate, transport, bioavailability, and toxicity in biosolids-amended soils for the targeted TOrCs.

Keywords: Trace organic chemicals, biosolids, fate and transport, terrestrial risk assessment.

TABLE OF CONTENTS

LIST OF TABLES

LIST OF FIGURES

LIST OF ACRONYMS

ADI	Acceptable daily intake
AHTN	Acetyl-hexamethyl--tetrahydronaphthalene
ARG	Antibiotic resistance gene
ATSDR	Agency for Toxic Substances and Disease Registry
BAF	Bioaccumulation factor
BFR	Brominated flame retardant
BDE	Brominated diphenyl ether
BOD	Biochemical oxygen demand
BPA	Bisphenol A
BSAF	Biota soil/sediment accumulation factor
bw	Body weight
CASRN	Chemical Abstracts Service Registry Number
CERHR	Center for the Evaluation of Risks to Human Reproduction
CFR	Code of Federal Regulations
CICADs	Concise International Chemical Assessment Documents
CIP	Ciprofloxacin
CTC	Chlortetracycline
DEHP	Di-(2-ethylhexyl) phthalate
DEP	Diethyl phthalate
DGGE	Denatured gradient gel electrophoresis
DOM	Dissolved organic matter
DNA	Deoxyribonulcelic acid
DSCADA	Disodium cocoamphodiacetate
DTC	Doxycycline
dw	Dry weight
EC_{10}	Effective concentration 10%
EC_{50}	Effective concentration 50%
ED_{10}	Effective dose 10%
ED_{50}	Effective dose 50%
ED_{90}	Effective dose 90%
EDC	Endocrine disrupting chemical
E1	Estrone
E2	17β-estradiol
E3	Estriol
EE2	17α-ethinyl estradiol
EU	European Union
EUSES	European Union System for the Evaluation of Substances
FDA	United States Food and Drug Administration
f_{oc}	Fraction organic carbon
FOSA	Perfluorooctane sulfonamide
FTOH	Fluorotelomer alcohol
HBCD	Hexabromocyclododecane
HERA	Human and Environmental Risk Assessment
HHCB	Hexahydro hexamethylcyclopentabenzopyran (Galaxolide)

IPCS	International Programme on Chemical Safety
IRIS	Integrated Risk Information System
JECFA	Joint FAO/WHO Expert Committee on Food Additives
K_d	Solid-water distribution coefficient
K_{dom}	Dissolved organic mater partitioning coefficient
K_{ow}	Octanol-water partitioning coefficient
K_{oc}	Organic-carbon normalized solid-water distribution coefficient
K_s	Half saturation biotransformation rate constant
LC_{50}	50% lethal concentration
LD_{50}	50% lethal dose
LOEC	Lowest observed effect level
MAPK	Mitogen activated protein kinase
MeEE2	Mestranol
mg/L	milligram per liter
N-EtFOSAA	2-*N*-ethylperfluorooctanesulfanamido acetic acid
ng/L	nanogram per liter
N-MeFOSAA	2-*N*-methylperfluorooctanesulfanamido acetic acid
NOEC	No observed effect concentration
NOEL	No observed effect level
NP	Nonylphenol
NP1EO	Nonylphenol monoethoxylate
NP2EO	Nonylphenol diethoxylate
OECD	Organization for Economic Cooperation and Development
OPPTS	Office of Pollution Prevention and Toxic Substances (U.S. EPA)
OTC	Oxytetracycline
PAH	Polycyclic aromatic hydrocarbon
PAPs	Polyfluoroalkyl phosphoric acids
PBDE	Polybrominated diphenyl ether
PCB	Polychlorinated biphenyl
PEC	Predicted environmental concentration
PFC	Perfluorochemical
PFDS	Perfluorodecane sulfonate
PFDA	Perfluorodecanoate
PFDoA	Perfluorododecanoate
PFHxS	Perfluorohexane sulfonate
PFNA	Perfluorononanoate
PFOS	Perfluorooctane sulfonate
PFOA	Perfluorooctanoate
PFTeDA	Perfluorotetradecanoate
PFUnDA	Perfluoroundecanoate
pH	Negative log of the hydronium ion concentration
pK_a	Negative log of the acid dissociation constant
PLFA	Phospholipids fatty acid analysis
PNEC	Predicted no effect concentration
PPAR	Peroxisome proliferator-actived receptor
PPCP	Pharmaceutical and personal care product
QSAR	Quantitative structure activity relationship

RCR	Risk characterization ratio
RfD	Reference dose
rRNA	Ribosomal ribonucleic acid
SDBS	Sodium dodecyl benzene sulfonate
SDS	Sodium dodecyl sulfate
SETAC	Society of Environmental Toxicology and Chemistry
S_w	Aqueous solubility
TBBPA	Tetrabromobisphenol A
TC	Tetracycline
TCC	Triclocarban
TCS	Triclosan
TDI	Tolerable daily intake
TGD	Technical Guidance Document
TNSSS	Targeted National Sewage Sludge Survey
TOrC	Trace organic chemical
U.S. EPA	U.S. Environmental Protection Agency
USGS	U.S. Geological Survey
WWTP	Wastewater treatment plant

EXECUTIVE SUMMARY

ES.1 Introduction

The land application of biosolids has a long history of demonstrated beneficial use and minimal environmental and human health risks when conducted in accordance with existing regulations. However, concerns posed by the presence of trace organic chemicals (TOrCs) in biosolids necessitate additional evaluation and risk assessment of the practice. The presence of TOrCs in biosolids does not necessarily mean there is cause for concern, as the risk of a TOrC will depend on both the exposure and effects to targeted receptors. Thus, the key question that needs to be addressed is - does the presence of TOrCs in biosolids pose a significant risk to ecological and human health following land application? This question can be addressed by conducting risk assessments for the chemicals of concern.

To help address this important question, the Water Environment Research Foundation (WERF) commissioned this study with the overall goals of: 1) identifying TOrCs of greatest potential concern; and 2) determining data gaps for conducting human and ecological risk assessments that could ultimately be used to support risk management decisions. An intended outcome of this study was to also help lay the groundwork for developing future research priorities. The scope of this project was focused on the terrestrial environment, in part due to resources allocated for the study and the interest of WERF Subscribers, though it should be recognized that aquatic and human exposure pathways may be important for some TOrCs.

ES.2 Objectives

This study was undertaken with three primary objectives in mind. First, an evaluation was made to determine the chemicals of greatest potential concern in the terrestrial environment. This evaluation led to two separate categories - high priority and low priority TOrCs. This assessment was based on occurrence data and readily available information on basic properties such as bioaccumulation and toxicity. Second, an evaluation of quantitative risk assessments was conducted to identify the most important parameters for conducting ecological risk assessments and the techniques currently available for obtaining the parameter values. A minimum data set for risk modeling was identified from this review. Third, a comprehensive literature review of the identified chemicals of greatest potential concern (high priority TOrCs) was conducted to identify relevant data on fate, transport, biotransfer from soil to plants and animals, and toxicity in the terrestrial environment. Based on the results from this review, data gaps were identified for the parameters most important for conducting terrestrial risk assessments.

ES.3 Research Approach

The study was initiated with a review of TOrCs reported to occur in municipal sewage sludge or biosolids. This list of TOrCs was then evaluated and prioritized, resulting in a list of high priority and low priority TOrCs for which additional data were sought and data gaps identified. Once the high priority list was identified, data on the occurrence, mobility, persistence, bioaccumulation, toxicity, and microbial impacts were sought for the targeted TOrCs, with a particular emphasis on data available with respect to biosolids-borne TOrCs in soils. The primary focus of the data search was on TOrCs identified as high priority. Congruently, a separate effort aimed at evaluating risk modeling approaches for biosolids-borne TOrCs was conducted. Results from the risk model evaluation were then used to identify the most critical parameters for conducting ecological risk assessments. Once the data were compiled and key parameters identified, the high priority TOrCs were evaluated for data gaps.

Finally, once data were compiled and data gaps identified for the high priority TOrCs, an assessment of relative data availability was conducted for classes (and, in some cases, subclasses) of TOrCs for each of the data types sought. For each category of data, each class or subclass of TOrCs was placed within one of four tiers (Tier 0 through Tier 3). The higher tier designation of relative data availability indicates that more data and/or higher quality data were available for that particular class or subclass of TOrCs. To enable this categorization of data availability, general criteria were developed specific to the type of data under evaluation. The goal of this effort was to provide an indication of where significant data gaps are with respect to the data requirements for risk modeling for each class or subclass of the high priority TOrCs.

ES.4 Prioritization

The final list of high and low priority TOrCs is presented in Table ES-1. High priority was assigned to TOrCs present at relatively high concentrations (> 1000 μg/kg) in biosolids as determined in one of two national surveys. In addition to the TOrCs with concentrations in biosolids > 1000 μg/kg, TOrCs identified as chemicals of particular concern in aquatic environments were also considered. Thus, the inclusion of TOrCs such as brominated flame retardants (BFRs), perfluorochemicals (PFCs), and synthetic steroidal chemicals as high priority TOrCs was based in part on expert judgment and public concern. The ultimate goal was to identify TOrCs of greatest potential concern to the environment and human health, but because of limited data and the approach taken for prioritization, the list may change as new data become available. Thus, the list of high priority TOrCs should be thought of as an evolving list, and chemicals can be added or deleted as new knowledge is gained.

Table ES-1. Trace Organic Chemicals Included in This Study.

Chemical(s)	CASRN	Chemical Class (Subclass)[a]	Use
High Priority			
BDE 28	41318-75-6	BFRs	Fire Retardant
BDE 47	5436-43-1	BFRs	Fire Retardant
BDE 85	182346-21-0	BFRs	Fire Retardant
BDE 99	60348-60-9	BFRs	Fire Retardant
BDE 100	189084-64-8	BFRs	Fire Retardant
BDE 138	182677-30-1	BFRs	Fire Retardant
BDE 153	68631-49-2	BFRs	Fire Retardant
BDE 154	207122-15-4	BFRs	Fire Retardant
BDE 183	207122-16-5	BFRs	Fire Retardant
BDE 209	1163-19-5	BFRs	Fire Retardant
Dimethyl TBBPA	37853-61-5	BFRs	Fire Retardant Metabolite
HBCD isomers	25637-99-4	BFRs	Fire Retardant
TBBPA	79-94-7	BFRs	Fire Retardant
10:2/12:2diPAPs	NA	PFCs and Precursors	Surface Coatings
10:2diPAPs	NA	PFCs and Precursors	Surface Coatings
6:2/8:2diPAPs	NA	PFCs and Precursors	Surface Coatings

[a] For the purposes of data gap analysis, PPCPs considered high priority were further subclassified depending on their uses.

Table ES-1. Trace Organic Chemicals Included in This Study (continued).

Chemical(s)	CASRN	Chemical Class (Subclass)[a]	Use
		High Priority (continued)	
6:2diPAPs	NA	PFCs and Precursors	Surface Coatings
8:2/10:2diPAPs	NA	PFCs and Precursors	Surface Coatings
8:2diPAPs	NA	PFCs and Precursors	Surface Coatings
FOSA	754-91-6	PFCs and Precursors	Surface Coatings
FOSAA	NA	PFCs and Precursors	Surface Coatings
N-EtFOSAA	NA	PFCs and Precursors	Surface Coatings
N-MeFOSAA	NA	PFCs and Precursors	Surface Coatings
PFDA	335-76-2	PFCs and Precursors	Surface Coatings
PFDoDA	307-55-1	PFCs and Precursors	Surface Coatings
PFDS	335-77-3	PFCs and Precursors	Surface Coatings
PFHpA	375-85-9	PFCs and Precursors	Surface Coatings
PFHxA	307-24-4	PFCs and Precursors	Surface Coatings
PFHxS	355-46-4	PFCs and Precursors	Surface Coatings
PFNA	375-95-1	PFCs and Precursors	Surface Coatings
PFOA	335-67-1	PFCs and Precursors	Surface Coatings
PFOS	1763-23-1	PFCs and Precursors	Surface Coatings
PFTeDA	376-06-7	PFCs and Precursors	Surface Coatings
PFTriDA	72629-94-8	PFCs and Precursors	Surface Coatings
PFUnDA	2058-94-8	PFCs and Precursors	Surface Coatings
Bisphenol A (BPA)	80-05-7	Plasticizers	Plasticizer
4-Epitetracycline	23313-80-6	PPCPs (Antibiotics)	Antibiotic
Ciprofloxacin (CIP)	85721-33-1	PPCPs (Antibiotics)	Antibiotic
Doxycycline	564-25-0	PPCPs (Antibiotics)	Antibiotic
Miconazole	22916-47-8	PPCPs (Antibiotics)	Antifungal
Ofloxacin	82419-36-1	PPCPs (Antibiotics)	Antibiotic
Tetracycline	60-54-8	PPCPs (Antibiotics)	Antibiotic
Triclocarban (TCC)	101-20-2	PPCPs (Antimicrobials)	Antimicrobial
Triclosan (TCS)	3380-34-5	PPCPs (Antimicrobials)	Antimicrobial
Galaxolide (HHCB)	80450-66-4	PPCPs (Musks)	Fragrance material
Tonalide (AHTN)	21145-77-7	PPCPs (Musks)	Fragrance material
Cimetidine	51481-61-9	PPCPs (Other)	Antacid
17α-Ethinyl estradiol (EE2)	57-63-6	Steroidal Chemicals	Synthetic hormone
Mestranol (MeEE2)	72-33-3	Steroidal Chemicals	Synthetic hormone
4-Cumylphenol	599-64-4	Surfactants	Detergent Metabolite
4-tert-octyl phenol	140-66-9	Surfactants	Detergent Metabolite
		Low Priority	
N-alkanes (polychlorinated)	NA	Aliphatics	Flame retardant
Polydimethylsiloxane (PDMS)	9016-00-6	Aliphatics	Organosilicone polymer
Polyorganosiloxanes	NA	Aliphatics	Organosilicone polymer
Propene (trichloro)	96-19-5	Aliphatics	Herbicide intermediate
Dibutyltin	1002-53-5	Organotins	Anti-fouling agent
Monobutyltin	2406-65-7	Organotins	Heat stabilizer/ anti-fouling agent
Tributyltin	688-73-3	Organotins	Anti-fouling Agent
Hexachlorophene (HCP)	70-30-4	Phenols	Disinfectant
Hydroquinone	123-31-9	Phenols	Photographic developing
Cresyldiphenyl phosphate	26444-49-5	Phosphate Esters	Plasticizer/flame retardant
Acetyl Cedrene	125783-65-5	PPCPs	Fragrance Material
Azithromycin	83905-01-5	PPCPs	Antibiotic
BLS	NA	PPCPs	Fluorescent whitening agent
DAS 1	16090-02-1	PPCPs	Fluorescent whitening agent
Diphenhydramine	58-73-1	PPCPs	Antihistamine
Diphenyl Ether	101-84-8	PPCPs	Fragrance material
DSBP	38775-22-3	PPCPs	Fluorescent whitening agent
Galaxolide lactone (HHCB-lactone)	NA	PPCPs	Fragrance material metabolite
Hexyl salicylate	6259-76-3	PPCPs	Fragrance material
Hexylcinnamic aldehyde (α)	101-86-0	PPCPs	Fragrance material
Ibuprofen	15687-27-1	PPCPs	Analgesic

[a] For the purposes of data gap analysis, PPCPs considered high priority were further subclassified depending on their uses.

Table ES-1. Trace Organic Chemicals Included in This Study (continued).

Chemical(s)	CASRN	Chemical Class (Subclass)[a]	Use
		Low Priority (continued)	
Iso-E-Super (OTNE)	54464-57-2	PPCPs	Fragrance material
Methyl ionone (gamma)	127-51-5	PPCPs	Fragrance material
Minocycline	10118-90-8	PPCPs	Antibiotic
Musk Ketone (MK)	81-14-1	PPCPs	Fragrance material
Phantolide (AHMI)	15323-35-0	PPCPs	Fragrance material
Sulfanilamide	63-74-1	PPCPs	Antibiotic
Thiabendazole	148-79-8	PPCPs	Anthelminitic
Traseolide (ATII)	68857-95-4	PPCPs	Fragrance material
17α-Dihydroequilin	651-55-8	Steroidal Chemicals	Steroid hormone
17α-Estradiol	57-91-0	Steroidal Chemicals	Steroid hormone
17β-Estradiol (E2)	50-28-2	Steroidal Chemicals	Steroid hormone
Androstenedione	63-05-8	Steroidal Chemicals	Steroid hormone
Androsterone	53-41-8	Steroidal Chemicals	Steroid hormone
Equilenin	517-09-9	Steroidal Chemicals	Steroid hormone
Equilin	474-86-2	Steroidal Chemicals	Steroid hormone
Estriol (E3)	50-27-1	Steroidal Chemicals	Steroid hormone
Estrone (E1)	53-16-7	Steroidal Chemicals	Steroid hormone
Etiocholanolone	53-42-9	Steroidal Chemicals	Androgen metabolite
Norethindrone	68-22-4	Steroidal Chemicals	Synthetic hormone
Norgestrel	6533-00-2	Steroidal Chemicals	Synthetic hormone
Progesterone	57-83-0	Steroidal Chemicals	Steroid hormone
Testosterone	58-22-0	Steroidal Chemicals	Steroid hormone
β-Estradiol-3-benzoate	50-50-0	Steroidal Chemicals	Synthetic hormone
$C_{10}EO_x$ (Alcohol Ethoxylates)	74432-13-6 (AEOs)	Surfactants	Surfactant
$C_{11}DEA$ (Coconut Diethanol Amide)	68603-42-9	Surfactants	Surfactant
$C_{12}EO_x$ (Alcohol Ethoxylates)	NA	Surfactants	Surfactant
$C_{13}DEA$ (Coconut Diethanol Amide)	68603-42-9	Surfactants	Surfactant
$C_{14}EO_x$ (Alcohol Ethoxylates)	68154-96-1 ($C_{14-18}EO_4$)	Surfactants	Surfactant
$C_{15}DEA$ (Coconut Diethanol Amide)	68603-42-9	Surfactants	Surfactant
$C_{16}EO_x$ (Alcohol Ethoxylates)	68154-96-1	Surfactants	Surfactant
$C_{17}DEA$ (Coconut Diethanol Amide)	68603-42-9	Surfactants	Surfactant
$C_{18}EOx$ (Alcohol Ethoxylates)	68154-96-1	Surfactants	Surfactant
Poly(ethylene glycol)s	25322-68-3	Surfactants	Polymer

[a] For the purposes of data gap analysis, PPCPs considered high priority were further subclassified depending on their uses.

ES.5 Risk Assessment Modeling

An evaluation of risk assessment models was conducted to identify: 1) parameters of most importance for conducting ecological risk assessments; 2) available methods for filling the data gaps; and 3) future needs for model improvements. An intent of this effort was to help guide the data gap analysis.

Risk assessment models are used to estimate contaminant exposure and inputs to the food chain transfer models. The transfer models typically include the uptake of TOrCs by plants grown in amended fields, accumulation by fruits and vegetables, and uptake by beef and dairy cattle that consume forage and silage grown on the biosolids-amended fields. Exposure estimates are also compared to the following ecological endpoints: 1) fish, aquatic invertebrates, aquatic plants, amphibians, aquatic community and sediment biota in the farm pond; 2) soil invertebrates and plants in the agricultural field; and 3) mammals and birds in contact with the agricultural field and farm pond. Due to the large number of potential receptors, an ecological effects assessment typically focuses on a small number of indicator organisms representative of the most exposed or the most sensitive species. Other data needed for risk assessment modeling include: chemical properties such as water solubility, vapor pressure, dissociation constants (pK_a), and octanol-water partitioning coefficients (K_{ow}; where appropriate); volatilization and degradation rates; organic carbon normalized solid-water partition coefficients ($K_{oc;}$ where appropriate) and soil-water partition coefficients (K_d); bioconcentration and bioaccumulation factors (BAFs) for ecological assessments; and biotransfer factors for human health assessments.

At present, all of the methods for predicting biouptake require a K_{ow} value. Most relationships between K_{ow} and biouptake were developed for hydrophobic organic chemicals, but many of the TOrCs included in the present study are not strongly hydrophobic. Various mathematical relationships have also been developed for predicting K_{oc} and K_d from a K_{ow} value, but these relationships are highly dependent on the chemical class (structural similarities). Terrestrial prey BAFs are generally not available and suitable relationships have not been established. Thus, a BAF of 1 is generally assumed for terrestrial prey, though higher BAFs may be possible if significant biomagnification occurs. In the absence of BAFs, small mammal BAFs are used for all terrestrial vertebrate prey and earthworm BAFs are used for all terrestrial invertebrate prey. Volatilization can be estimated from chemical properties using the U.S. EPA EPIWIN computer program. Water solubility and vapor pressure values can also be used to predict the Henry's Law constant. Degradation rates (biodegradation, hydrolysis, and photolysis) are difficult to predict and thus are best measured. If no empirical values are available, a default value of zero (no degradation) is assumed.

The minimum data set required by U.S. EPA to begin evaluating risk associated with contaminants in biosolids is presented in Table ES-2. A rigorous sensitivity analysis is needed to quantify the effect of changes in model parameters on the model outcome. For the minimum data set parameters, values should only be used if: 1) they are produced using accepted and appropriate analytical techniques, published in peer reviewed studies, or reports; or 2) they can be appropriately estimated using U.S. EPA-approved or other peer reviewed methods. The minimum data set could be refined by establishing a screening model methodology focusing on a specific subset of exposure pathways considered relevant to the type of chemical being evaluated (e.g., bioaccumulative chemicals). However, such a methodology does not currently exist and, therefore, the minimum data set focuses on parameters currently required to run the risk assessment model. The table also includes references to *in silico* models (e.g., EPI Suite, SPARC) that could be used to estimate model parameters. Further investigation is needed, though, to determine whether these estimation techniques could be used for the TOrCs identified in this review. Direct measurement of these parameters is preferred, thus acceptable test methods are also included in the table.

Risk assessment modeling is an iterative process. As new knowledge is obtained, the risk assessment assumptions and model formulations need to be reevaluated. Based on our current knowledge of the fate and transport of TOrCs in biosolids-amended soil, several new model formulations are proposed to better describe these processes. These new model formulations include better descriptions of: 1) the sorption of ionogenic TOrCs in soil; 2) kinetically controlled sorption of TOrCs in soil; 3) the sorption of TOrCs to colloidal material involved in facilitated transport; 4) kinetic degradation of TOrCs beyond a simple first-order loss in soil; and 5) the incorporation of biotransformation of TOrCs in plants. Perhaps most importantly, the current risk models need to be verified with field validation studies. This verification exercise would include models for predicting biosolids concentrations as well as exposure concentrations in the terrestrial environment.

Table ES-2. Minimum Data Set Required for the U.S. EPA Risk Assessment.

Parameter	Module(s)	Test Methods	Estimation Techniques
Health benchmark	Human risk	Cancer potency factors, ingestion reference doses	Surrogate chemical or most toxic chemical in class
Ecological benchmark	Ecological risk	Water quality criteria, soil quality criteria, lowest affect dose for population endpoint	Surrogate chemical or estimation programs like ECOSAR for aquatic life
Molecular weight (and chemical structure)	Source, Surface water	-	None
Partition coefficients	Multiple	OPPTS 835.1220 (OECD 106) OPPTS 830.7550 (OECD 107)	EPI Suite, SPARC, and established estimation equations
Water Solubility	Source, Water modules	OPPTS 830.7840 (OECD 105)	EPI Suite, SPARC
Critical pressure	Source	-	EPI Suite, SPARC
Critical temperature	Source	-	EPI Suite
Boiling point	Source	OPPTS 830.7220 (OECD 103)	EPI Suite, SPARC
Vapor pressure coefficients	Source	OPPTS 830.7950 (OECD 104)	EPI Suite, SPARC
Henry's Law Constant	Multiple	-	EPI Suite and established estimation equations
Diffusivity in air	Source	-	SPARC
Diffusion coefficient in water	Source, Groundwater	-	SPARC
Ionization equilibrium constant (requires acid base designation)	Multiple	OPPTS 830.7370 (OECD 112)	SPARC
Soil degradation rate*	Watershed, Source	OPPTS 835.3110 (OECD 301)	EPI Suite
Surface water degradation rate*	Surface water	OPPTS 835.4100 (OECD 307)	EPI Suite
Groundwater degradation rate*	Groundwater	OPPTS 835.6100	EPI Suite
Bioconcentration factors	Aquatic food web	OPPTS 850.1850 OPPTS 850.1730 (OECD 305)	EPI Suite
Bioaccumulation factors	Terrestrial food web	OPPTS 850.4800 OPPTS 850.6200 (OECD 207)	Methods available for plants and worms, but not well developed for other prey
Biotransfer factors	Farm food chain	OPPTS 870.8320 OPPTS 870.8340	Available methods for plant uptake, beef, and dairy

* An overall media-specific degradation rate is typically a function of degradation rate associated with specific biotic and abiotic processes such as biodegradation (aerobic and anaerobic), hydrolysis, photolysis, etc. Note that for screening purposes, degradation rates are sometimes assumed to be zero or very low (using a highly persistent organic chemical as a surrogate) to support a conservative model simulation.

ES.6 Occurrence of Trace Organic Chemicals in Municipal Biosolids

The process used to identify and prioritize TOrCs for consideration relied heavily on detection and quantitation in municipal sewage sludge or biosolids, and substantial occurrence data were available for nearly all of the TOrCs targeted in this study. However, for some TOrCs, notably the perfluorochemicals (PFCs) and PFC precursors, a substantial occurrence data base is lacking. The two national surveys employed as primary sources of occurrence data (the U.S. Geological Survey (USGS) survey and the U.S. EPA's Targeted National Sewage Sludge Survey (TNSSS)) were fairly comprehensive. The TNSSS had a substantial sample size, whereas the USGS survey was more limited in scope. Broad surveys, such as the TNSSS, are needed to ensure that representative concentrations are used in risk assessments.

During the initial data collection effort, it became apparent that this study would be limited by what has already been detected in biosolids. While this is a necessary limitation of the scope of the study, it suggests that an effort to identify TOrCs that *might potentially occur* in biosolids (and therefore *might potentially pose a risk* to humans and the environment) is needed. Such a WERF-sponsored effort was recently conducted for household chemicals in wastewater (03CTS21UR), and a similar effort is currently underway with respect to assessing impacts of wastewater treatment plant effluents on receiving water bodies. An effort specific to biosolids may identify TOrCs for which analytical methods should be developed. The current study is limited by what has already been measured, rather than considering what should be measured.

Data on the occurrence of biosolids-borne TOrCs in biosolids-amended soils were also sought. Not surprisingly, several of the targeted TOrCs were identified and quantified in soils amended with biosolids (either as part of an experimental plot or through normal agricultural practices). However, many of the identified studies lacked clear data as to when biosolids were applied and/or the concentrations of the TOrCs present in the applied biosolids. Such data are *crucial* for interpretation of the levels observed, and thus there is a clear need for additional controlled field studies. Studies of this type would address the persistence and mobility of biosolids-borne TOrCs, rather than simply occurrence. These studies should employ sufficient replicates and sampling frequencies to enable meaningful interpretation of trends observed over time.

ES.7 Mobility of Biosolids-Borne Trace Organic Chemicals in Soils

Considerable data were available with respect to understanding the potential mobility of the targeted TOrCs in biosolids-amended soils. In particular, physicochemical parameters such as soil-water partition coefficients and octanol-water partition coefficients were available for many of the targeted TOrCs. However, some of the TOrCs examined in this study (i.e., tetracycline, ciprofloxacin, and perfluorochemicals) do not follow the traditional hydrophobic organic contaminant paradigm. Clearly, *appropriate* applications of existing modeling approaches and alternative modeling approaches are needed to adequately describe the mobility behavior of TOrCs in biosolids-amended soils.

Understanding the mobility of biosolids-borne TOrCs in biosolids-amended soils requires working beyond the laboratory-scale, and evaluating mobility in bench-scale column studies and, especially, in field-scale experiments. Unfortunately, few such studies exist for several of the targeted TOrCs. Some studies indicated that some TOrCs can leach from fields, particularly when the applied biosolids are not dewatered, whereas other TOrCs (e.g., polybrominated diphenyl ethers, synthetic musks, and some steroidal chemicals) exhibited low leaching potential. More comprehensive bench and field-scale studies (with respect to analytes) are needed to accurately represent the real-world conditions under which biosolids are applied.

Bench and field-scale experiments on TOrC mobility would also help address another major data gap identified for nearly all of the TOrCs. The issue of irreversible sorption (chemisorption) and desorption of the targeted TOrCs from soils was not addressed in most of the published mobility studies. Biosolids present a unique matrix in soils, and for the same reason that many TOrCs may not fit the traditional partitioning paradigms (the presence of active functional groups), there is likely a greater potential for these TOrCs to become irreversibly bound to either soil organic matter or the biosolids-derived organic matter. Conversely, the binding of deprotonated TOrCs (many of the targeted TOrCs exist as anions at environmentally relevant pH values) can be substantially less than predicted from the hydrophobicity of the neutral form. In general, neither pH-dependent sorption nor the potential for irreversible (or strongly hysteretic) sorption are considered in traditional mobility models, many of which assume reversible sorption. Further studies of desorption in all compound classes are required to identify the TOrCs for which this is an important issue.

ES.8 Persistence of Biosolids-Borne Trace Organic Chemicals in Soils

The persistence of biosolids-borne TOrCs in soils is a result of many processes, but biodegradation is generally considered the dominant process affecting TOrC attenuation in biosolids-amended soils. For most of the high priority TOrCs, no soil biodegradation data were identified from the peer-reviewed literature. In the absence of these data, it may be possible to make qualitative predictions of biodegradability for some TOrCs based on data from analogous chemicals. In particular, while certain pharmaceuticals, personal care product ingredients, and steroidal chemicals have benefited from research in both aquatic and soil systems, others lack such data in environmental systems most relevant to biosolids amendment. This deficit was true for all brominated flame retardants (BFRs) as well as many of the perfluorochemical precursors. When soil or biosolids-specific transformation data were unavailable, data from aquatic systems were considered as general indicators of recalcitrance, though their applicability to biosolids-amended soils is tenuous.

Environmental factors such as pH, moisture content, metal cations, temperature, and bacterial cell concentration all can affect biodegradation rates. The effects of such factors and the impact of different soil types or biosolids loading rates on attenuation need to be further investigated for most targeted TOrCs. Literature for antimicrobials and antibiotics indicate recalcitrance and slow biodegradation in soil systems and a dependence on site characteristics such as biosolids content, aerobic conditions, and soil depth. Biodegradation rates of steroidal chemicals can be favorably impacted by the presence of biosolids, increased temperatures, and adequate (but not excessive) water content in soils. Unfortunately, degradation data for many of the TOrCs included in this study are lacking for soils and biosolids-amended soils. Hence, discerning the current rate-limiting TOrCs that could mandate solids loading or biosolids application rates is difficult. Most TOrCs are transformed to less toxic intermediates, but aqueous studies of some of the targeted TOrCs highlight the potential for more toxic degradation products, particularly for the polybrominated diphenyl ethers and perfluorochemical precursors. Whether these processes also occur in soil systems remains unclear. Indeed, the behavior of likely degradation products of target compounds is little studied and deserves research attention.

Future TOrC biodegradation research should focus on soil and biosolids-amended soil systems to better understand the risks associated with biosolids-borne TOrCs in the environment. Data pertaining to brominated flame retardants, perfluorochemical precursors, plasticizers, and surfactants would benefit most from additional biosolids-focused research, but so would most of the TOrCs targeted in this analysis.

ES.9 Bioaccumulation of Biosolids-Borne Trace Organic Chemicals in Soils

Data on the bioaccumulation of biosolids-borne TOrCs in plants and animals were examined, but few useful data sets were found. Bioaccumulation of some of the TOrCs has been documented, but few studies examined bioaccumulation and bioavailability specifically in biosolids-amended soils. Since data derived from biosolids-amended systems were extremely limited, general accumulation data from soils was also evaluated. Bioaccumulation data from sediments (particularly for animals) were also compiled, but the relevance of these studies to biosolids-amended soils is questionable.

Some of the targeted TOrCs (tetracycline antibiotics, antimicrobials, fluoroquinolones, and synthetic musks, brominated flame retardants) can accumulate in a variety of plants including grass, green onions, cabbage, corn, alfalfa, lettuce, radish, zucchini, and carrots. Data for other compound classes were generally absent. More data were available for the bioaccumulation and bioavailability of TOrCs in animals, particularly invertebrates such as earthworms. Unfortunately, many of the studies identified did not provide significant detail as to the exposure conditions, making the modeling of the bioaccumulation highly problematic. Parameters such as BAFs and biota-soil accumulation factors (BSAFs) are meant to facilitate comparisons of bioavailability between chemicals and between sites, but, factor units were not always provided or were inconsistent, making comparisons difficult.

Some TOrCs such as perfluorochemicals do not accumulate in lipids, rendering lipid normalization inappropriate. In addition, the affinity of other TOrCs for the solid phase (i.e., tetracyclines) does not necessarily depend on organic carbon, rendering organic carbon normalization problematic. These factors point to a need for consistency in measuring and reporting data to facilitate comparisons among TOrCs. Factors that should be considered include the organisms used (i.e., standard organisms), how the chemical is introduced to the organism, use of environmentally relevant conditions, and standardization of units and methods of normalization to calculate BAF and BSAF values.

The bioaccumulation data evaluated in this study focused on the uptake of TOrCs from biosolids amended soils to either plants or animals. While *biomagnification*, a process by which the body burden of the TOrC increases as the trophic level increases, of TOrCs has not been demonstrated from biosolids-amended soil, it may occur for some TOrCs depending on their chemical properties and the food chain pathway. Consideration of such processes may be important for risk models, and at least one recent study suggested trophic transfer may be the most sensitive pathway for risks associated with biosolids-borne triclocarban. However, this assessment was based on model predictions and thus would need to be verified by laboratory or field experiments, and such data are extremely limited.

ES.10 Toxicity of Biosolids-Borne Trace Organic Chemicals in Soils

Even though the focus of this study was on the terrestrial environment, both human and ecological toxicity data were sought for the high priority TOrCs. Only publicly-available toxicity data were sought. Relevant human toxicity values were identified for less than half of the targeted TOrCs, though an exhaustive review may have identified more human toxicity data. Furthermore, the data gathered should be further scrutinized with regard to the confidence they engender. For example, substantial bodies of data and expert scientific review were involved in the development of toxicity values for perfluorochemicals and bisphenol A (BPA), but little chemical-specific information was available for development of the 4-cumylphenol threshold of toxicological concern. For some TOrCs the mode of action, particularly if additive toxicity is possible, should be evaluated in more detail.

Ecotoxicological data for the targeted TOrCs were sought, but relevant soil and sediment toxicity data were found for only a few of the high priority TOrCs. Even when relevant ecotoxicity studies were identified, they were limited in terms of quantity, study quality, toxicological endpoints investigated, and number of species and taxa evaluated. A significant proportion of the available studies were conducted in sediment, and the applicability of these studies for soils, much less biosolids-amended soils, is highly questionable. Significant data gaps exist with respect to the toxicity of the targeted TOrCs in terrestrial environments, particularly in biosolids-amended soils. Comparatively, there are substantial volumes of toxicity data for many of the targeted TOrCs in aquatic environments, and WERF-sponsored evaluations of these data are currently underway. However, while some of the exposure pathways relevant for biosolids-amended soils are aquatic in nature, aquatic toxicity data were not sought in this review (with the noted exception of sediment studies). Even including the aquatic toxicity data, the known and potential modes of action of the targeted TOrCs in ecological receptors are far from comprehensive and should be expanded in future efforts. Clearly, additional studies are needed to examine the terrestrial toxicity of the targeted TOrCs in biosolids-amended soils, particularly studies that include trophic transfer and toxicity to higher trophic level organisms.

ES.11 Impacts on Soil Microbial Communities

Though biosolids are most often land applied in agricultural settings where land management practices are expected to alter natural microbial ecosystems, some TOrCs may have toxicological effects on soil macrobiota or soil microbial communities over and above the effects of the biosolids. Microbial impacts can be measured by examining alterations in community composition, metabolic function, and diversity. Data pertaining to the microbial impacts, much less microbial impacts in biosolids-amended soils, of many of the targeted TOrCs were not available. Where possible, data for analogous chemicals within the identified classes were evaluated. Studies identified for the targeted TOrCs demonstrate a variety of effects on soil microbial structure and function, though data derived from biosolids-amended soils were limited. The types of effects observed include suppression of soil nitrification rates, increases in antibiotic resistance, and other general changes to community structure, metabolism, and diversity. However, few generalizations can be made, even within specific classes of TOrCs. Exposure of microbial communities to some pharmaceuticals and personal care product ingredients increased microbial biomass, richness, and the sizes of certain bacterial populations, whereas others increased the presence of antibiotic resistance genes (ARGs).

A better understanding of potential selective pressures resulting from the introduction of the TOrCs in a biosolids matrix is needed to adequately address the risk and acceptable loads associated with the land application of biosolids. This includes consortia structural properties, propagation of ARGs, and functional processes such as attenuation and nutrient cycling. Few data are available for many of the targeted TOrCs, suggesting a significant need for research into the potential effects of the targeted TOrC on microbial systems in managed soils.

ES.12 Overall Data Gaps

A summary of data availability for the high priority TOrCs is presented in Table ES-3. Substantial data gaps exist for many processes important to understanding the risk of biosolids-borne TOrCs in soil environments. Some data are available for particular processes (e.g., sorption), but few data were found specifically with respect to biosolids-amended soils. Very few studies were identified that were intentionally designed to address the fate, transport, bioaccumulation, and toxicity of TOrCs in biosolids-amended soils under well-controlled conditions. The complexity of the biosolids matrix is often ignored in many studies, and the potential for irreversibly bound residues of TOrCs in biosolids-amended soils has not appropriately modeled, much less adequately characterized. Bench-scale column studies may be appropriate avenues of research for addressing specific questions related to the fate and transport of biosolids-borne TOrCs in soils. *The most significant data gap, however, is the absence of human toxicological and ecotoxicological data as well as biotransfer data for ecological receptors.* Some of these data may be available from industry studies, and, if made available, this should be considered a potential source oftoxicity data for futureefforts. *In addition, well-studied long-term field plots to which biosolids have been applied would also help in exposure and effects evaluations, as well as validation of the risk model predictions.* Such field studies can incorporate the complexities often ignored in more controlled laboratory settings, and offer the possibility of conducting studies with multiple objectives (i.e., mobility, bioaccumulation, and volatilization) under identical conditions. Additional data are needed to reduce the uncertainty in assessing the risk of biosolids-borne TOrCs in the terrestrial environment.

Table ES-3. Summary of Data Availability for the High Priority Trace Organic Chemicals.

Chemical Class	Occurrence	Mobility	Persistence	Bio-availability	Toxicity		Microbial Impacts
					Human	Ecological	
Brominated Flame Retardants(BFRs)	Tier 3	Tier 1	Tier 1	Tier 2	Tier 0	Tier 0	Tier 0
Perfluorochemicals (PFCs) and PFC Precursors	Tier 1	Tier 2	Tier 1	Tier 0	Tier 0	Tier 0	Tier 0
PPCPs: Antimicrobials	Tier 3	Tier 2	Tier 3	Tier 1	Tier 0	Tier 0	Tier 1
PPCPs: Antibiotics	Tier 3	Tier 2	Tier 1	Tier 0	Tier 2	Tier 0	Tier 1
PPCPs: Musks	Tier 3	Tier 2	Tier 3	Tier 2	Tier 1	Tier 0	Tier 0
PPCPs: Other	Tier 3	Tier 0	Tier 0	Tier 0	Tier 2	Tier 0	Tier 0
Plasticizers	Tier 3	Tier 2	Tier 1	Tier 1	Tier 1	Tier 0	Tier 0
Steroidal Chemicals	Tier 3	Tier 2	Tier 2	Tier 1	Tier 2	Tier 0	Tier 0
Surfactants	Tier 3	Tier 2	Tier 0	Tier 1	Tier 1	Tier 0	Tier 0

Generic Interpretation of Data Availability:

Tier 0	Essentially no data were available of this type for this class or subclass of TOrCs, including data that could be used for modeling.
Tier 1	For the majority of TOrCs in this class or subclass, some data were available, but available data are likely of limited utility or are limited to modeled systems only (i.e., not directly derived from experimental studies).
Tier 2	Useful data from experimental systems are available for a majority of TOrCs in this class or subclass, but most of the data are not directly applicable to biosolids-amended soils.
Tier 3	Substantial data of this type directly relevant for biosolids-amended soils are available, though some gaps in data may exist for specific TOrCs. For this class or subclass of TOrCs, data are available that have been measured in real world systems with biosolids-borne TOrCs and reasonable biosolids application rates, and/or in long-term field-based studies with appropriate attention to study design and QA/QC.

Though a detailed chemical-by-chemical analysis of the data gaps relevant to modeling the risk of biosolids-borne TOrCs in soils was beyond the scope of the present effort, Table ES-3 provides a compound-class assessment of data gaps for the targeted TOrCs. For each class or subclass of TOrCs and for each type of data sought as part of this study, a ranking assessment of data availability was made using a four-tier system (Tier 0 = essentially no data to Tier 3, substantial data). The specific criteria used for this assessment are provided in the chapters of this report, but general descriptors of data availability are provided as a footnote to the table. These criteria were not meant to be definitive nor directly comparable between the data types: a Tier 2 ranking with respect to persistence does not necessarily indicate the same quality nor quantity of data are available when compared to a Tier 2 ranking with respect to mobility. Rather, this table is meant to present the relative data availability within each class, and to emphasize the types of and relative extent of data gaps for the high priority TOrCs examined in this study. Moreover, expert judgment was often required to assign the ranking, as compiled data did not always neatly fit within the ranking criteria frameworks developed for each type of data.

ES.13 Future Research Agenda

The purpose of this research effort was to identify data gaps with respect to understanding and predicting the potential risk of biosolids-borne TOrCs in biosolids-amended soils. Based on the findings from this study, suggestions for possible next steps are provided below. The intent of these suggestions is to help lay the groundwork for developing future research priorities.

In that vein, suggestions for future research are:

1.) Conduct a preliminary risk analysis on the high priority TOrCs, using data compiled in this effort and expert judgment, to identify those compounds with the highest probability of causing harm to the environment and human health. This analysis could lead to a top 10 list of TOrCs or TOrC-classes of potential concern.

2.) Conduct comprehensive laboratory studies of appropriate quality to fill in the data gaps for the TOrCs or TOrC classes of greatest potential concern (subset derived from the top 10 list). These studies would improve and refine the risk assessments for the highest priority TOrCs, and would be aimed at raising the data availability for these TOrCs to at least "Tier 2" for all categories of data tabulated in Table ES-3.

3.) Conduct field studies to verify the risk model calculations and to provide insight on improving the risk model formulations. This effort would be aimed at raising the data availability for these TOrCs to "Tier 3" for all relevant categories of data tabulated in Table ES-3.

4.) Evaluate current risk models for their ability to protect the environment and human health. Quantification of uncertainties associated with the input parameters would be determined.

5.) Conduct a sensitivity analysis on the current risk models to identify which parameters are the most important in predicting risk. This effort would also help quantify the uncertainty associated with model input parameters and thus could be combined with an overall effort to assess risk models for their ability to protect the environment and human health.

6.) Develop a screening level risk model that can be used to assess TOrCs with limited data. Such a model could be used to identify TOrCs of immediate concern. This effort could be combined with conducting a preliminary risk analysis of the high priority TOrCs.

7.) Develop a tiered data selection hierarchy that could be used to characterize the uncertainty of input parameters in the risk estimates.

8.) Develop analytical methods for measuring TOrCs of concern in complex matrices (e.g., biosolids and soils).

CHAPTER 1.0

INTRODUCTION

1.1 Historical Context

The presence of trace organic chemicals (TOrCs) in municipal biosolids destined for land application has received increasing attention by the public and regulatory community in recent years. However, while concerns about so-called "emerging" organic contaminants in biosolids, such as many pharmaceutical and personal care products (PPCPs) or perfluorochemicals (PFCs), have become more pressing, many of these TOrCs have likely been present in biosolids for decades. It is only now, with the advent of more sensitive analytical techniques and toxicological assays that our understanding of the occurrence and potential effects of TOrCs is beginning to develop. The main objective of this study is to review existing literature and determine data gaps that limit our ability to assess potential adverse environmental and human health impacts of biosolids-borne TOrCs in soils.

The land application of biosolids has a long history of demonstrated beneficial use and minimal environmental and human health risks when conducted in accordance with existing regulations. However, concerns about the presence of TOrCs in municipal biosolids are not new. Indeed, various organic chemicals were included in the initial listing of 200 "pollutants of concern" in sewage sludge in 1984 (U.S. EPA, 1995). Included in this list were many priority pollutants. In subsequent years and through a series of four independent review panels, this list was reduced to 50 TOrCs prior to undergoing "worst case" hazard index assessments that included 14 separate pathways for adverse human health and environmental impacts. While limits were initially identified for 13 organic chemicals based on these detailed risk assessments, following the completion of the 1990 National Sewage Sludge Survey, organic chemicals were deleted from the promulgated 40 CFR Part 503 Rule because:

- the use of the chemical was already banned, or it was no longer manufactured; or
- the chemical occurred at insignificant concentration in biosolids; or
- the chemical concentrations in biosolids was lower than the limits proposed as a result of the detailed risk assessments.

In subsequent years, substantial efforts were made with respect to collecting data on the occurrence and potential risks associated with biosolids-borne TOrCs. Research efforts were expanded to include chemicals beyond the initial 13 TOrCs of concern, and the Round Two National Sewage Sludge Survey was conducted to acquire additional occurrence data on TOrCs, specifically dioxins and dioxin-like chemicals (U.S. EPA, 1996). Using data generated from this survey, a sophisticated and thorough probabilistic risk assessment was applied to biosolids-borne dioxins and dioxin-like compounds (e.g., dioxins, furans, polychlorinated biphenyls (PCBs)). This led the U.S. Environmental Protection Agency (U.S. EPA) to decide that "no numeric limits or management practices are required to adequately protect human health and the environment from the adverse health effects of dioxins in land-applied sewage sludge." In essence, U.S. EPA concluded that the improvements to human health and environmental health would be trivial if limits on the presence of these TOrCs in biosolids were set (U.S. EPA, 2003a).

Some claim that accurate risk assessments for TOrCs in biosolids destined for land application have yet to be conducted, but the presence of TOrCs in biosolids and in biosolids-amended soils has received significant attention by the scientific and regulatory community for decades. To date, most research has suggested that risks to human health are minimal (NRC, 2002), though as suggested earlier, our understanding of the some of the potential risks associated with biosolids-borne TOrCs is itself only beginning to emerge. Further, a serious concern among the public and scientific community is the recurring lack of sufficient data to accurately assess potential ecological impacts of biosolids-borne chemicals. Particularly in light of the large diversity of biologically-active TOrCs (such as PPCPs) recently detected in biosolids (Kinney et al., 2006; U.S. EPA, 2009i), careful analysis of what is known (and more importantly still unknown) about the potential risks associated with biosolids-borne TOrCs is warranted.

1.2 Chemicals of Emerging Concern

A full decade has passed since the publication of Christian Daughton's and Thomas Ternes' report on PPCPs in the environment (Daughton and Ternes, 1999). Since then, many other TOrCs have been added to the list of "chemicals of emerging concern." Special issues of various journals have been dedicated to the topic (e.g., January 2005 issue of *Environmental Pollution*, December 2006 issue of *Environmental Science & Technology*). Interest in TOrCs of emerging concern in biosolids has also benefited from advances in detection and quantification techniques. For example, stable isotope dilution methods coupled to mass spectrometry have permitted lower and lower limits of detection and the minimization of the analytical effects associated with the inherently complex biosolids matrix.

Of particular concern for some of these TOrCs is their design to be biologically active at extremely low concentrations and, thus, to potentially exert adverse effects via exposure pathways not considered in traditional risk assessments. However, it is unclear to what extent these effects might materialize in biosolids-amended soils. Concerns remain about the potential impact of biosolids-borne TOrCs on terrestrial ecosystems and the potential for biosolids-borne TOrCs to leach into potable water supplies. As many TOrCs of emerging concern are considerably less hydrophobic than the organic chemicals previously considered by the U.S. EPA, contamination of water resources may be a more important pathway than previously recognized. However, the less hydrophobic a chemical, the less likely it will partition to and accumulate in biosolids in the first place.

1.3 Purpose and Scope of the Present Study

To help address the question of whether the presence of TOrCs in biosolids pose a significant risk to ecological and human health following land application, the Water Environment Research Foundation (WERF) commissioned this study with the overall goals of: 1) identifying TOrCs of greatest concern and, 2) determining data gaps for conducting ecological and human health risk assessments critical to risk management decisions. An intended outcome of this study was to also help lay the groundwork for developing future research priorities. The scope of this project wasfocused on the terrestrial environment, in part, due to resources allocated for the study and the interest of WERF Subscribers, though it should be recognized that aquatic and human exposure pathways may be important for some TOrCs.

To address these goals, the completion of three primary objectives was sought. The first objective was to determine the TOrCs of greatest concern with respect to the land application of biosolids. A second objective was to evaluate the various quantitative risk assessment approaches applicable to biosolids-borne TOrCs and identify the most important parameters for conducting ecological risk assessments and the techniques currently available for obtaining the parameter values. The third objective was to conduct a comprehensive literature review of the identified TOrCs of greatest concern and identify relevant data on fate, transport, biotransfer from soil to plants and animals, and toxicity in the terrestrial environment with the end goal of identifying the scientific data gaps for the parameters most important for conducting terrestrial risk assessments.

1.3.1 Research Approach

The first step in the data gap analysis was to identify the TOrCs for which data gaps should be evaluated. This prioritization is described in Chapter 2.0 and included substantial collection and verification of occurrence data from a variety of governmental reports and peer-reviewed studies. The result was a list of targeted TOrCs for which additional data were sought. The primary focus was on TOrCs identified as "high" priority TOrCs, though data on the "low" priority TOrCs were also collected and evaluated when readily available. A discussion of the approach used to differentiate between high and low priority TOrCs is included in Chapter 2.0.

Next a review of current methodologies for assessing risk is presented in Chapter 3.0. The review illustrates the type of data needed for conducting a terrestrial risk assessment and appropriate models. A key outcome of this exercise was identification of the minimum data set (i.e., parameters) that is needed for each TOrC to conduct meaningful terrestrial risk assessments.

Following prioritization, the review of risk methodology, and identification of the minimum data set needed for risk assessment modeling, additional data were collected and analyzed for the target TOrCs specifically related to:

- Occurrence
- Mobility
- Persistence
- Bioavailability
- Toxicity
- Effects on Soil Microbial Ecosystems

Each topic is discussed in separate chapters of this report. A brief overview of each chapter is provided below.

Risk Assessment Modeling (Chapter 3.0)

Risk assessment models rely on a wealth of information to assess the exposure and effects of TOrCs in the environment. This includes information on a chemical's fate (biotransformation, sorption and volatilization) and transport in the terrestrial environment to assess exposure, as well as potential effects and bioaccumulation as a result of exposure. In lieu of laboratory and field studies, various computer models have been used to predict physical and chemical properties of TOrCs that are then used to estimate their fate, effects and bioaccumulation in the terrestrial environment. Chapter 3.0 provides a review and evaluation of modeling approaches currently used in the United States and Europe. This evaluation was conducted to identify: 1) parameters of most importance for conducting ecological risk assessments, 2) available methods for filling the data gaps, and 3) future needs for model improvements. An intent of this effort was to help guide the data gap analysis.

Occurrence (Chapter 4.0)

This chapter includes data on the occurrence of the target TOrCs in sewage sludge and biosolids and, where data are available, in biosolids-amended soils. For both the high priority and low priority TOrCs, occurrence data in sewage sludge or biosolids are tabulated. For TOrCs that were excluded from this study based upon infrequent detection or low occurrence, occurrence data are provided in Chapter 2.0 (Prioritization).

Mobility (Chapter 5.0)

This chapter summarizes studies that have evaluated the tendency of TOrCs to move in the environment, including data on the potential for TOrCs to reach groundwater. Data collected and evaluated include studies on physical transport (i.e., runoff, volatilization), leaching to groundwater, and sorption. A tabulation of various measured and modeled physicochemical parameters that could be used to predict the sorption (and hence mobility) of biosolids-borne TOrCs is provided. Also provided is a brief discussion of the relevant sorption mechanisms likely controlling the mobility of each class or subclass of targeted TOrCs.

Persistence (Chapter 6.0)

Studies examining the processes relating to the persistence of TOrCs in soils are collected and analyzed in Chapter 6.0. As most attenuation processes are microbially-driven, the primary focus was to collect data on biosolids-borne TOrC biotransformation and biodegradation. To the extent data were available for soils (whether or not they were amended with biosolids), biodegradation data were sought. However, in the absence of soils data, a summary of aqueous phase biotransformation studies was also included. Data on abiotic transformations (e.g., photolysis, hydrolysis) were not specifically sought as these processes were presumed to be of secondary importance for most of the TOrCs included in this study.

Bioavailability (Chapter 7.0)

Data on the bioavailability and bioaccumulation of biosolids-borne TOrCs in soils are evaluated in Chapter 7.0. Limited data examining the bioaccumulation of biosolids-borne TOrCs by both plants and animals (e.g., earthworms) were available for analysis.

Toxicity (Chapter 8.0)

Separate from efforts to understand the bioaccumulative potential of biosolids-borne TOrCs in soils, Chapter 8.0 focuses on analyzing studies on the adverse (i.e., toxic) effects of TOrCs on organisms. Depending on the availability and applicability of data, general toxicity data for the select TOrCs were included in this analysis. Only summary data (e.g., reference doses) pertinent to human toxicological endpoints were collected, as an extensive evaluation of human toxicological data was beyond the scope of the study. The chapter focuses on soil benchmark concentrations or soil quality guidelines established for the target TOrCs.

Effects on Soil Microbial Ecosystems (Chapter 9.0)

This chapter evaluates data on the potential effects of TOrCs on microbially-driven processes in soils. Data on the impacts of TOrCs on soil microbe dynamics, including community changes and antibacterial resistance are collected and summarized.

1.4 Data Gap Analysis

Though a detailed chemical-by-chemical analysis of the data gaps relevant to modeling the risk of biosolids-borne TOrCs in soils was beyond the scope of the present effort, an analysis of the relative availability of the each type of data collected and summarized in Chapters 4.0 through 9.0 was conducted. For each class or subclass of TOrCs and for each type of data sought as part of this study, a ranking assessment of data availability was made using a four-tier system (Tier 0 = essentially no data to Tier 3, substantial data). The specific criteria used for this assessment are provided at the end of the respective chapters. These criteria were not meant to be definitive nor directly comparable between the data types: a Tier 2 ranking with respect to persistence does not necessarily indicate the same quality nor quantity of data are available when compared to a Tier 2 ranking with respect to mobility. Rather, the analysis presents the relative data availability within each class, and emphasizes the types and relative extent of data gaps for the high priority TOrCs. Expert judgment was often required to assign the rankings, as compiled data rarely fit neatly within the ranking criteria frameworks developed for each type of data.

CHAPTER 2.0

SELECTION AND PRIORITIZATION OF TARGET TRACE ORGANIC CHEMICALS

2.1 Selection Process for Biosolids-Borne Trace Organic Chemicals

There are thousands of synthetic organic chemicals that have the potential to occur in municipal biosolids and adversely affect environmental and human health (Harrison et al., 2006; Smith, 2008; Hydromantis, 2009b). A comprehensive review of the state of the science with respect to all chemicals potentially present in municipal biosolids was beyond the scope of the current study. Rather, the focus of this effort was to identify data gaps for *emerging* trace organic chemicals (TOrCs) that have not been as widely studied or evaluated with respect to their presence in biosolids or possible impacts. Thus, we conducted an initial screening of the TOrCs for which data gaps should be identified (described below). The result of the screening effort is the list of targeted chemicals for which data were collected and data gaps assessed (Table 2-1).

To ensure adequate resources were allocated to identifying data gaps, the list of chemicals were divided into two tiers of priority: high and low. A brief discussion of the process used for prioritization is also included below. Due to time and budgetary constraints, we focused on identifying data gaps primarily for the high priority TOrCs. These chemicals are the focus of this report, and additional data on their occurrence, mobility, persistence, bioavailability, toxicity, and microbial impacts can be found in Chapters 4.0 through 9.0, respectively.

Table 2-1. Trace Organic Chemicals Included in this Study.

Chemical(s)	CASRN	Chemical Class (Subclass)	Use
High Priority			
BDE 28	41318-75-6	BFRs	Fire Retardant
BDE 47	5436-43-1	BFRs	Fire Retardant
BDE 85	182346-21-0	BFRs	Fire Retardant
BDE 99	60348-60-9	BFRs	Fire Retardant
BDE 100	189084-64-8	BFRs	Fire Retardant
BDE 138	182677-30-1	BFRs	Fire Retardant
BDE 153	68631-49-2	BFRs	Fire Retardant
BDE 154	207122-15-4	BFRs	Fire Retardant
BDE 183	207122-16-5	BFRs	Fire Retardant
BDE 209	1163-19-5	BFRs	Fire Retardant
Dimethyl TBBPA	37853-61-5	BFRs	Fire Retardant Metabolite
Hexabromocyclododecane (HBCD) isomers	25637-99-4	BFRs	Fire Retardant
TBBPA	79-94-7	BFRs	Fire Retardant
10:2/12:2diPAPs		PFCs and Precursors	Surface Coatings
10:2diPAPs		PFCs and Precursors	Surface Coatings
6:2/8:2diPAPs		PFCs and Precursors	Surface Coatings
6:2diPAPs		PFCs and Precursors	Surface Coatings
8:2/10:2diPAPs		PFCs and Precursors	Surface Coatings
8:2diPAPs		PFCs and Precursors	Surface Coatings

Table 2-1. Trace Organic Chemicals Included in this Study (continued).

Chemical(s)	CASRN	Chemical Class (Subclass)	Use
High Priority (continued)			
FOSA	754-91-6	PFCs and Precursors	Surface Coatings
FOSAA	NA	PFCs and Precursors	Surface Coatings
N-EtFOSAA	NA	PFCs and Precursors	Surface Coatings
N-MeFOSAA	NA	PFCs and Precursors	Surface Coatings
PFDA	335-76-2	PFCs and Precursors	Surface Coatings
PFDoDA	307-55-1	PFCs and Precursors	Surface Coatings
PFDS	335-77-3	PFCs and Precursors	Surface Coatings
PFHpA	375-85-9	PFCs and Precursors	Surface Coatings
PFHxA	307-24-4	PFCs and Precursors	Surface Coatings
PFHxS	355-46-4	PFCs and Precursors	Surface Coatings
PFNA	375-95-1	PFCs and Precursors	Surface Coatings
PFOA	335-67-1	PFCs and Precursors	Surface Coatings
PFOS	1763-23-1	PFCs and Precursors	Surface Coatings
PFTeDA	376-06-7	PFCs and Precursors	Surface Coatings
PFTriDA	72629-94-8	PFCs and Precursors	Surface Coatings
PFUnDA	2058-94-8	PFCs and Precursors	Surface Coatings
Bisphenol A (BPA)	80-05-7	Plasticizers	Plasticizer
4-Epitetracycline	23313-80-6	PPCPs (Antibiotics)	Antibiotic
Ciprofloxacin (CIP)	85721-33-1	PPCPs (Antibiotics)	Antibiotic
Doxycycline (DTC)	564-25-0	PPCPs (Antibiotics)	Antibiotic
Miconazole	22916-47-8	PPCPs (Antibiotics)	Antifungal
Ofloxacin	82419-36-1	PPCPs (Antibiotics)	Antibiotic
Tetracycline (TC)	60-54-8	PPCPs (Antibiotics)	Antibiotic
Triclocarban (TCC)	101-20-2	PPCPs (Antimicrobials)	Antimicrobial
Triclosan (TCS)	3380-34-5	PPCPs (Antimicrobials)	Antimicrobial
Galaxolide (HHCB)	80450-66-4	PPCPs (Musks)	Fragrance material
Tonalide (AHTN)	21145-77-7	PPCPs (Musks)	Fragrance material
Cimetidine	51481-61-9	PPCPs (Other)	Antacid
17α-Ethinyl estradiol (EE2)	57-63-6	Steroidal Chemicals	Synthetic hormone
Mestranol (MeEE2)	72-33-3	Steroidal Chemicals	Synthetic hormone
4-Cumylphenol	599-64-4	Surfactants	Detergent Metabolite
4-tert-octyl phenol	140-66-9	Surfactants	Detergent Metabolite
Low Priority			
N-alkanes (polychlorinated)	Mixed	Aliphatics	Flame retardant
Polydimethylsiloxane (PDMS)	9016-00-6	Aliphatics	Organosilicone polymer
Polyorganosiloxanes	Mixed	Aliphatics	Organosilicone polymer
Propene (trichloro)	96-19-5	Aliphatics	Herbicide intermediate
Dibutyltin	1002-53-5	Organotins	Anti-fouling agent
Monobutyltin	2406-65-7	Organotins	Heat stabilizer/ anti-fouling agent
Tributyltin	688-73-3	Organotins	Anti-fouling Agent
Hexachlorophene (HCP)	70-30-4	Phenols	Disinfectant
Hydroquinone	123-31-9	Phenols	Photographic developing
Cresyldiphenyl phosphate	26444-49-5	Phosphate Esters	Plasticizer/flame retardant
Acetyl Cedrene	125783-65-5	PPCPs	Fragrance Material
Azithromycin	83905-01-5	PPCPs	Antibiotic
BLS		PPCPs	Fluorescent whitening agent
DAS 1	16090-02-1	PPCPs	Fluorescent whitening agent
Diphenhydramine	58-73-1	PPCPs	Antihistamine
Diphenyl Ether	101-84-8	PPCPs	Fragrance material
DSBP	38775-22-3	PPCPs	Fluorescent whitening agent
Galaxolide lactone (HHCB-lactone)	NA	PPCPs	Fragrance material metabolite
Hexyl salicylate	6259-76-3	PPCPs	Fragrance material
Hexylcinnamic aldehyde (α)	101-86-0	PPCPs	Fragrance material
Ibuprofen	15687-27-1	PPCPs	Analgesic
Iso-E-Super (OTNE)	54464-57-2	PPCPs	Fragrance material
Methyl ionone (gamma)	127-51-5	PPCPs	Fragrance material
Minocycline	10118-90-8	PPCPs	Antibiotic
Musk Ketone (MK)	81-14-1	PPCPs	Fragrance material

Table 2-1. Trace Organic Chemicals Included in this Study (continued).

Chemical(s)	CASRN	Chemical Class (Subclass)	Use
Low Priority (continued)			
Phantolide (AHMI)	15323-35-0	PPCPs	Fragrance material
Sulfanilamide	63-74-1	PPCPs	Antibiotic
Thiabendazole	148-79-8	PPCPs	Anthelminitic
Traseolide (ATII)	68857-95-4	PPCPs	Fragrance material
17α-Dihydroequilin	651-55-8	Steroidal Chemicals	Steroid hormone
17α-Estradiol	57-91-0	Steroidal Chemicals	Steroid hormone
17β-Estradiol (E2)	50-28-2	Steroidal Chemicals	Steroid hormone
Androstenedione	63-05-8	Steroidal Chemicals	Steroid hormone
Androsterone	53-41-8	Steroidal Chemicals	Steroid hormone
Equilenin	517-09-9	Steroidal Chemicals	Steroid hormone
Equilin	474-86-2	Steroidal Chemicals	Steroid hormone
Estriol (E3)	50-27-1	Steroidal Chemicals	Steroid hormone
Estrone (E1)	53-16-7	Steroidal Chemicals	Steroid hormone
Etiocholanolone	53-42-9	Steroidal Chemicals	Androgen metabolite
Norethindrone	68-22-4	Steroidal Chemicals	Synthetic hormone
Norgestrel	6533-00-2	Steroidal Chemicals	Synthetic hormone
Progesterone	57-83-0	Steroidal Chemicals	Steroid hormone
Testosterone	58-22-0	Steroidal Chemicals	Steroid hormone
β-Estradiol-3-benzoate	50-50-0	Steroidal Chemicals	Synthetic hormone
$C_{10}EO_x$ (Alcohol Ethoxylates)	74432-13-6 (AEOs)	Surfactants	Surfactant
$C_{11}DEA$ (Coconut Diethanol Amide)	68603-42-9	Surfactants	Surfactant
$C_{12}EO_x$ (Alcohol Ethoxylates)		Surfactants	Surfactant
$C_{13}DEA$ (Coconut Diethanol Amide)	68603-42-9	Surfactants	Surfactant
$C_{14}EO_x$ (Alcohol Ethoxylates)	68154-96-1 (C14-18EO4)	Surfactants	Surfactant
$C_{15}DEA$ (Coconut Diethanol Amide)	68603-42-9	Surfactants	Surfactant
$C_{16}EO_x$ (Alcohol Ethoxylates)	68154-96-1	Surfactants	Surfactant
$C_{17}DEA$ (Coconut Diethanol Amide)	68603-42-9	Surfactants	Surfactant
$C_{18}EOx$ (Alcohol Ethoxylates)	68154-96-1	Surfactants	Surfactant
Poly(ethylene glycol)s	25322-68-3	Surfactants	Polymer

2.2 Selection Approach

2.2.1 Categorical Exclusions

As a first step in the screening process, chemicals were excluded for which substantial data are available and for which risk assessments and/or regulatory standards have already been developed. Thus, chemicals identified as priority pollutants (U.S. EPA, 2009h), pesticides, or belonging to another class of well-studied chemicals (i.e., polycyclic aromatic hydrocarbons; PAHs), were excluded. The list of chemicals that were excluded on this basis is provided in Table 2-2.

2.2.2 Regulated Elsewhere or Previously Evaluated

As a second step in narrowing the list of target TOrCs, chemicals were also excluded for which regulatory standards have been enacted related to their presence in biosolids (Table 2-2). The rationale for excluding these chemicals was that if regulatory decisions have already been made, substantial data regarding potential risk to human health and the environment likely exist. Smith (2009) discusses the rationale behind the regulations and differences between regulatory frameworks with respect to chemicals in biosolids. Some of the regulations are largely numerically based (as opposed to risk-based) and there may be substantial data gaps required for complete risk assessments. The focus of the current effort is on TOrCs for which risk assessments and regulations might be considered.

Table 2-2. Trace Organic Chemicals Excluded from Data Gap Analysis.

Chemical	Chemical Class	Reason for Exclusion
1,1,1- Trichloroethane	Aliphatics	Previously Evaluated[1]
1-2-Dichloroethane	Aliphatics	Previously Evaluated[1]
1-methylnaphthalene	PAHs	PAH
2,6-dimethylnaphthalene	PAHs	PAH
2-methylnaphthalene	PAHs	PAH
4-Chloroaniline	Cyclics	Previously Evaluated[3]
4-Chloro-3-methylphenol	Phenols	Previously Evaluated[1]
4-nonylphenol (NP)	Surfactants	Regulated Elsewhere[4]
Acenaphthene	PAHs	Priority Pollutant[2]
Acetophenone	Cyclics	Previously Evaluated[3]
Acrylonitrile	Aliphatics	Priority Pollutant[2]
Aldrin	Pesticides	Priority Pollutant[2]
Aniline (2,4,5-trimethyl)	Cyclics	Previously Evaluated[1]
Anthracene	PAHs	Priority Pollutant[2]
Azinphos Methyl	Pesticides	Previously Evaluated[3]
Benzene	Cyclics	Priority Pollutant[2]
Benzene (1,4-dinitro)	Cyclics	Previously Evaluated[1]
Benzene (dichloro) isomers	Chlorobenzenes	Priority Pollutant[2]
Benzene (ethyl)	Cyclics	Priority Pollutant[2]
Benzene (hexachloro)	Chlorobenzenes	Priority Pollutant[2]
Benzene (monochloro)	Chlorobenzenes	Priority Pollutant[2]
Benzene (mononitro)	Cyclics	Priority Pollutant[2]
Benzene (pentachloro)	Chlorobenzenes	Previously Evaluated[1]
Benzene (pentachloronitro)	Pesticides	Pesticide
Benzene (tetrachloro)	Chlorobenzenes	Previously Evaluated[1]
Benzene (trichloro) isomers	Chlorobenzenes	Priority Pollutant[2]
Benzenethiazole (2-methylthio)	Cyclics	Previously Evaluated[1]
Benzenethiol	Cyclics	Previously Evaluated[1]
Benzidine	PAHs	Priority Pollutant[2]
Benzo(a)anthracene	PAHs	Priority Pollutant[2]
Benzo(a)pyrene	PAHs	Previously Evaluated[3]
Benzo(b)fluoranthene + Benzo(k)fluoranthene	PAHs	Priority Pollutant[2]
Benzo(g,h.i)perylene	PAHs	Priority Pollutant[2]
Benzofluorene congeners	PAHs	PAH
Benzoic acid	Cyclics	Previously Evaluated[3]
Benzopyrene congeners	PAHs	Priority Pollutant[2]
Benzyl alcohol	Cyclics	Previously Evaluated[1]
Biphenyl	PAHs	Previously Evaluated[3]
Biphenyl (decachloro)	PCBs/Dioxins	Previously Evaluated[3]
Bis (2-Ethylhexyl) phthalate (DEHP)	Plasticizers	Priority Pollutant[2]
Bis(2-chloroethoxy)methane	Plasticizers	Priority Pollutant[2]
Bis(2-chloroethyl)ether	Plasticizers	Priority Pollutant[2]
Bis(2-chloroisopropyl)ether	Plasticizers	Priority Pollutant[2]
Butane (1,2,3,4-diepoxy)	Aliphatics	Previously Evaluated[1]
Butanol (iso)	Aliphatics	Previously Evaluated[1]
Butanone (2-)	Aliphatics	Previously Evaluated[3]
Butylbenzyl phthalate (BBP)	Plasticizers	Priority Pollutant[2]
Campesterol	Steroidal Chemicals	Natural Source
Captan	Pesticides	Pesticide
Carbon disulfide	Aliphatics	Previously Evaluated[3]
Chlordane	Pesticides	Priority Pollutant[2]
Chlorobenzilate	Pesticides	Previously Evaluated[3]
Chloropyrifos	Pesticides	Previously Evaluated[3]
Chlorpyrifos	Pesticides	Pesticide
Cholestanol (5a-)	Steroidal Chemicals	Natural Source

Notes:
[1] Considered by U.S. EPA in the Round Two National Sewage Sludge Survey (U.S. EPA, 1996)
[2] Identified as a Priority Pollutant (http://www.epa.gov/waterscience/methods/pollutants.htm)
[3] Considered by U.S. EPA(U.S. EPA, 1992)
[4] Regulated in biosolids elsewhere (Smith, 2009)

Table 2-2. Trace Organic Chemicals Excluded from Data Gap Analysis (continued).

Chemical	Chemical Class	Reason for Exclusion
Cholesterol	Steroidal Chemicals	Natural Source
Chrysene	PAHs	Priority Pollutant[2]
Chrysene + triphenylene	PAHs	Priority Pollutant[2]
Ciodrin	Pesticides	Pesticide
Coprostanol	Steroidal Chemicals	Natural Source
Crotonaldehyde	Aliphatics	Previously Evaluated[1]
Cyclohexane isomers (lindane and others)	Pesticides	Priority Pollutant[2]
Cyclopentadiene (hexachloro)	Aliphatics	Previously Evaluated[1]
Cymene (P-)	Cyclics	Previously Evaluated[1]
DDT and related cogeners	Pesticides	Priority Pollutant[2]
Desmosterol	Steroidal Chemicals	Natural Source
Diallate	Pesticides	Pesticide
Diazinon	Pesticides	Previously Evaluated[3]
Dibenzo(ah)anthracene	PAHs	PAH
Dibenzoanthracene congeners	PAHs	Priority Pollutant[2]
Dibenzofuran	PCBs/Dioxins	PCBs/Dioxin
Dibenzothiophene	PAHs	PAH
Dicrotophos (Bidrin)	Pesticides	Pesticide
Dieldrin	Pesticides	Priority Pollutant[2]
Diethyl phthalate	Plasticizers	Priority Pollutant[2]
Dimethoate	Pesticides	Pesticide
Dimethyl phthalate (DMP)	Plasticizers	Priority Pollutant[2]
Di-n-butyl phthalate (DBP)	Plasticizers	Priority Pollutant[2]
Di-n-octyl phthalate (DnOP)	Plasticizers	Priority Pollutant[2]
Dioxane	Cyclics	Previously Evaluated[3]
Dioxins and furans (polychlorinated dibenzo)	PCBs/Dioxins	Priority Pollutant[2]
Diphenyl amine	PAHs	PAH
Disulfotone	Pesticides	Pesticide
Diuron	Pesticides	Pesticide
Endosulfan	Pesticides	Priority Pollutant[2]
Endrin	Pesticides	Priority Pollutant[2]
Epi-coprostanol	Steroidal Chemicals	Natural Source
Ergosterol	Steroidal Chemicals	Natural Source
Ethane (hexachloro)	Aliphatics	Priority Pollutant[2]
Ethane (monochloro)	Aliphatics	Priority Pollutant[2]
Ethane (pentachloro)	Aliphatics	Previously Evaluated[1]
Ethane (tetrachloro)	Aliphatics	Previously Evaluated[1]
Ethane (trichloro) isomers	Aliphatics	Priority Pollutant[2]
Ethylene (dichloro)	Aliphatics	Priority Pollutant[2]
Ethylene (monochloro)	Aliphatics	Priority Pollutant[2]
Ethylene (tetrachloro)	Aliphatics	Priority Pollutant[2]
Ethylene (trichloro)	Aliphatics	Priority Pollutant[2]
Famphur	Pesticides	Pesticide
Fluoranthene	PAHs	Priority Pollutant[2]
Fluorene	PAHs	Priority Pollutant[2]
Fluorene (nitro)	PAHs	PAH
Heptachlor	Pesticides	Priority Pollutant[2]
Heptachlor epoxides	Pesticides	Priority Pollutant[2]
Hexanoic acid	Aliphatics	Previously Evaluated[1]
Hexanone (2-)	Aliphatics	Previously Evaluated[1]
Indeno(1,2,3-cd)pyrene	PAHs	Priority Pollutant[2]
Indole	PPCPs	Natural Source
Isobenzan	Pesticides	Pesticide
Isodrin	Pesticides	Pesticide

Notes:
[1] Considered by U.S. EPA in the Round Two National Sewage Sludge Survey (U.S. EPA, 1996)
[2] Identified as a Priority Pollutant (http://www.epa.gov/waterscience/methods/pollutants.htm)
[3] Considered by U.S. EPA(U.S. EPA, 1992)
[4] Regulated in biosolids elsewhere (Smith, 2009)

Table 2-2. Trace Organic Chemicals Excluded from Data Gap Analysis (continued).

Chemical	Chemical Class	Reason for Exclusion
Isophorone	Pesticides	Priority Pollutant[2]
Leptophos	Pesticides	Pesticide
Linear alkylbenzene sulfonates (LAS)	Surfactants	Regulated Elsewhere[4]
Methane (dichloro)	Aliphatics	Priority Pollutant[2]
Methane (monochloro)	Aliphatics	Priority Pollutant[2]
Methane (tetrachloro)	Aliphatics	Priority Pollutant[2]
Methane (trichloro)	Aliphatics	Priority Pollutant[2]
Methane (trichlorofluoro)	Aliphatics	Previously Evaluated[3]
Methoxychlor	Pesticides	Pesticide
Methylnaphthalene isomers	PAHs	PAH
Methylphenanthrene isomers	PAHs	PAH
Mevinphos (phosdrin)	Pesticides	Pesticide
Naled (Dibrom)	Pesticides	Previously Evaluated[3]
N-alkanes	Aliphatics	Previously Evaluated[1]
Naphthalene	PAHs	Priority Pollutant[2]
Naphthalene nitro congeners	PAHs	PAH
Naphthoquinone (1,4-)	Pesticides	Pesticide
Nitrofen	Pesticides	Pesticide
N-nitrosdimethylamine	Nitrosamines	Previously Evaluated[3]
N-nitrosdiphenylamine	Nitrosamines	Priority Pollutant[2]
N-nitrosodiethylamine	Nitrosamines	Previously Evaluated[1]
N-nitrosodi-*n*-butylamine	Nitrosamines	Previously Evaluated[1]
N-nitrosomorpholine	Nitrosamines	Previously Evaluated[1]
N-nitrosopiperdine	Nitrosamines	Previously Evaluated[1]
NP tri-EO (NP3EO)	Surfactants	Regulated Elsewhere[4]
NP di-EO (NP2EO)	Surfactants	Regulated Elsewhere[4]
NP mono-EO (NP1EO)	Surfactants	Regulated Elsewhere[4]
Para-cresol	Phenols	Previously Evaluated[1]
Parathion (ethyl)	Pesticides	Pesticide
Parathion (methyl)	Pesticides	Pesticide
PCB congeners	PCBs/Dioxins	PCBs/Dioxins
Pentachlorophenol (PCP)	Phenols	Priority Pollutant[2]
Pentanone (methyl)	Aliphatics	Previously Evaluated[1]
Peremethrin isomers	Pesticides	Pesticide
Perylene	PAHs	PAH
Phenanthrene	PAHs	Priority Pollutant[2]
Phenol	Phenols	Priority Pollutant[2]
Phenol chloro congeners	Phenols	Priority Pollutant[2]
Phenol methyl congeners	Phenols	Priority Pollutant[2]
Phenol nitro methyl congeners	Phenols	Priority Pollutant[2]
Phenols nitro congeners	Phenols	Priority Pollutant[2]
Phenoxy herbicides	Pesticides	Pesticide
Phenoxypropanoic acid (trichloro)	Pesticides	Pesticide
Phenylether (chloro)	PCBs/Dioxins	Priority Pollutant[2]
Phorate	Pesticides	Pesticide
Phosphamidon	Pesticides	Pesticide
Picoline (2-)	Cyclics	Previously Evaluated[1]
Pronamide	Pesticides	Pesticide
Propane (dichloro) isomers	Aliphatics	Priority Pollutant[2]
Propane (trichloro)	Aliphatics	Previously Evaluated[1]
Propanenitrile (ethyl cyanide)	Aliphatics	Previously Evaluated[1]
Propanone (2-)	Aliphatics	Previously Evaluated[1]
Propen-1-ol (2-)	Aliphatics	Previously Evaluated[1]
Propene chlorinated isomers	Aliphatics	Previously Evaluated[1]

Notes:

[1] Considered by U.S. EPA in the Round Two National Sewage Sludge Survey (U.S. EPA, 1996)

[2] Identified as a Priority Pollutant (http://www.epa.gov/waterscience/methods/pollutants.htm)

[3] Considered by U.S. EPA(U.S. EPA, 1992)

[4] Regulated in biosolids elsewhere (Smith, 2009)

Table 2-2. Trace Organic Chemicals Excluded from Data Gap Analysis (continued).

Chemical	Chemical Class	Reason for Exclusion
Propenenitrile (methyl)	Aliphatics	Previously Evaluated[1]
Pyrene	PAHs	Priority Pollutant[2]
Pyrene (phenyl)	PAHs	PAH
Pyrophosphate (tetraethyl)	Pesticides	Pesticide
Quintozene	Pesticides	Pesticide
Retene (7-isopropyl-1-methylphenanthrene)	PAHs	PAH
Safrol (iso)	Pesticides	Pesticide
Safrole (EPN)	Pesticides	Previously Evaluated[3]
Skatol	PPCPs	Natural Source
Squalene	Aliphatics	Previously Evaluated[1]
Stigmasterol	Steroidal Chemicals	Natural Source
Styrene	Cyclics	Previously Evaluated[1]
Sulfone (dimethyl)	Aliphatics	Previously Evaluated[1]
Terpeniol	Cyclics	Previously Evaluated[1]
Terphenyls and naphthalenes (polychlorinated)	PCBs/Dioxins	Priority Pollutant[2]
Thioxanthe-9-one	Cyclics	Previously Evaluated[1]
Toluene	Cyclics	Priority Pollutant[2]
Toluene (dinitro)	Cyclics	Priority Pollutant[2]
Toxaphene	Pesticides	Priority Pollutant[2]
Trichlofon	Pesticides	Pesticide
Trichlorophenols	Phenols	Priority Pollutant[2]
Tricresyl phosphate	Phosphate Esters	Previously Evaluated[3]
Trifluralin (Treflan)	Pesticides	Previously Evaluated[3]
Triphenylene	PAHs	PAH
Xylene isomers	Cyclics	Previously Evaluated[3]
β-Sitosterol	Steroidal Chemicals	Natural Source
β-Stigmastanol	Steroidal Chemicals	Natural Source
Higher NP-EOs NP(4-17)EO	Surfactants	Regulated Elsewhere[4]
Butadiene (hexachloro-1,3)	Aliphatics	Priority Pollutant[2]
Campestanol (5a+5b)	Steroidal Chemicals	Natural Source
Sitostanol (5α-β+5β-β)	Steroidal Chemicals	Natural Source
Aroclor 1016	PCBs/Dioxins	Priority Pollutant[2]
Aroclor 1248	PCBs/Dioxins	Priority Pollutant[2]
Aroclor 1254	PCBs/Dioxins	Priority Pollutant[2]
Aroclor 1260	PCBs/Dioxins	Priority Pollutant[2]

Notes:

[1] Considered by U.S. EPA in the Round Two National Sewage Sludge Survey (U.S. EPA, 1996)

[2] Identified as a Priority Pollutant (http://www.epa.gov/waterscience/methods/pollutants.htm)

[3] Considered by U.S. EPA(U.S. EPA, 1992)

[4] Regulated in biosolids elsewhere (Smith, 2009)

Chemicals previously considered and/or evaluated by the U.S. EPA with respect to potential regulation in biosolids were also excluded from this study. The group includes all chemicals evaluated as part of the Round Two National Sewage Sludge Survey (U.S. EPA, 1996). As the objective of this study was to identify data gaps for TOrCs that have not previously been considered, the identification of data gaps for chemicals previously addressed by U.S. EPA was outside the scope of our effort.

2.2.3 Exclusions Based on Natural Occurrence

Numerous chemicals identified and quantified in biosolids are naturally produced biomolecules such as phytosterols. Other biomolecules (indole and skatole) are used in consumer products, but are also naturally occurring. These chemicals are often included in biosolids surveys as indicators of anthropogenic waste (Kinney et al., 2006) despite their natural occurrence. The risks associated with naturally-occurring TOrCs in biosolids are typically perceived as low, so they were excluded from the data gap analysis (Table 2-2) unless there was a known toxicological concern. Some biomolecules, such as the steroid hormones, were included in the data gap analysis despite their natural sources and low occurrence because of toxicological concerns. The rationale for this inclusion is discussed in section 2.2.5.

2.2.4 Exclusions Based on Occurrence Data

Of the remaining chemicals reported as possibly occurring in biosolids, many were either not detected in national surveys of biosolids, were detected at low frequency (< 10%), or were determined to have low concentrations in biosolids (maximum mean or median concentration less than 500 μg/kg and a maximum concentration less than 1000 μg/kg). No or low frequencies of detection for TOrCs in biosolids suggests low and/or highly variable sources, possibly resulting from isolated influences of industrial sources to municipal biosolids. Low frequency of detection does not indicate an absence of risk, but suggests that meaningful risk reduction would not likely be achieved through national standards regulating the levels of these chemicals in biosolids. Unless included for other reasons (section 2.2.5), TOrCs detected at high frequency, but at low levels (<1000 μg/kg) [1] were also excluded from the study. Significant dilution (i.e., 100-200 fold) of biosolids-borne TOrCs is likely to occur upon land application of biosolids at agronomic rates, resulting in very low levels in soils. Even for very persistent and bioaccumulative polychlorinated biphenyls (PCBs), current regulations in other countries only regulate the sum of several congeners in biosolids at the 200 to 800 μg/kg level (Smith, 2009). While a lower cutoff for the maximum average concentration would be more conservative, the focus of this effort was to identify chemicals whose presence may be most problematic. Generally (but not always) this corresponds to chemicals present at the highest concentrations. Chemicals that were excluded from this study on the basis of occurrence data are listed in Table 2-3, along with the occurrence data used to exclude them from the study.

Table 2-3. Trace Organic Chemicals Excluded from Data Gap Analysis Based on Occurrence Data.

Chemical	Chemical Class	Reason for Exclusion	Max Mean or Median (μg/kg)	Concentration Range (μg/kg)	Data Sources
1,7-Dimethylxanthine	PPCPs	< 10 % Detects	769	ND – 9,580	1, 2
2-Ethyl-hexanal	Plasticizers	Low Concentration	NA	ND	3
2-Ethyl-hexanoic acid	Plasticizers	Low Concentration	NA	ND	3
2-Ethyl-hexanol	Plasticizers	Low Concentration	NA	ND	3
3,4-Dichlorophenyl isocyanate	PPCPs	Low Concentration	165	ND – 530	1
4-Epianhydrochlortetracycline	PPCPs	< 10 % Detects	419	NA	2
4-Epianhydrotetracycline	PPCPs	Low Concentration	236	ND – 2,160	2
4-Epichlortetracycline	PPCPs	< 10 % Detects	118	ND – 974	2
4-Epioxytetracycline	PPCPs	Low Concentration	45	36 – 55	2
Acetaminophen	PPCPs	< 10 % Detects	454	ND – 1,300	2
Albuterol	PPCPs	< 10 % Detects	168	ND – 850	1

NA = not available; ND = not detected

Data Sources:

1 (Kinney et al., 2006)
2 (U.S. EPA, 2009i)
3 (Barnabe et al., 2008)
4 (Herren and Berset, 2000)
5 (Gielen, 2007)
6 (Jones-Lepp and Stevens, 2007)
7 (Kaleta et al., 2006)
8 (Petrovic and Barcelo, 2000)
9 (Kupper et al., 2004)
10 (Torslov et al., 1997)
11 (Radjenovic et al., 2009)
12 (Fent, 1989)
13 (Lee and Peart, 2002)
14 (Bolz et al., 2001)
15 (Smyth et al., 2007)
16 (Berset et al., 2000)
17 (Mumma et al., 1984)
18 (Spongberg and Witter, 2008)
19 (Xu et al., 2007)

[1] When average or median concentrations for a given chemical were available from multiple studies, the maximum value was used to determine its inclusion in this study. The exceptions were if the average data were from either the U.S. EPA Targeted National Sewage Sludge Survey (U.S. EPA, 2009i) or the USGS biosolids survey (Kinney et al., 2006): when available, the highest value from these two data sets were used regardless of other data sources.

Table 2-3. Trace Organic Chemicals Excluded from Data Gap Analysis Based on Occurrence Data (continued).

Chemical	Chemical Class	Reason for Exclusion	Max Mean or Median (µg/kg)	Concentration Range (µg/kg)	Data Sources
Amino Musk Ketone	PPCPs	Low Concentration	NA	ND – 13	4
Amino Musk Xylene	PPCPs	Low Concentration	13	ND – 49	4
Amitriptyline	PPCPs	Low Concentration	ND	ND	5
Amphetamine	PPCPs	Low Concentration	NA	5 – 300	6
Anhydrochlortetracycline	PPCPs	< 10 % Detects	124	ND – 125	2
Anhydrotetracycline	PPCPs	Low Concentration	246	ND – 1,960	2
Anthraquinone	PPCPs	Low Concentration	NA	ND – 217	1
Atenolol	PPCPs	Low Concentration	3	ND – 5	7
Benzophenone	PPCPs	Low Concentration	270	ND – 584	1
Bezafibrate	PPCPs	Low Concentration	70	ND – 88	7
C_7DEA (Coconut Diethanol Amide)	Surfactants	Low Concentration	NA	ND	8
C_9DEA (Coconut Diethanol Amide)	Surfactants	Low Concentration	NA	50 – 200	8
Caffeine	PPCPs	Low Concentration	224	9 – 1,010	1, 2
Carbadox	PPCPs	< 10 % Detects	NA	ND	2
Carbamazepine	PPCPs	Low Concentration	135	ND – 6,030	1, 2
Cashmeran (DPMI)	PPCPs	Low Concentration	93	38 – 332	5
Cefotaxime	PPCPs	< 10 % Detects	102	ND	2
Celestolide (ADBI)	PPCPs	Low Concentration	400	100 – 1,010	9
Chlorpromazine	PPCPs	Low Concentration	NA	ND	4
Chlortetracycline	PPCPs	< 10 % Detects	54	ND – 1,010	2
Clarithromycin	PPCPs	Low Concentration	40	ND – 617	2
Clinafloxacin	PPCPs	< 10 % Detects	NA	ND	2
Clofibric Acid	PPCPs	Low Concentration	34	ND – 64	7
Cloxacillin	PPCPs	< 10 % Detects	NA	ND	2
Codeine	PPCPs	Low Concentration	30	ND – 328	1, 2
Cotinine	PPCPs	Low Concentration	54	ND – 690	1, 2
d-limonene	PPCPs	Low Concentration	297	ND – 1,070	1
Dehydronifedipine	PPCPs	Low Concentration	5	ND – 26	1, 2
Demeclocycline	PPCPs	< 10 % Detects	105	ND – 200	2
Di (2-ethylhexyl) adipate	Plasticizers	Low Concentration	139	ND – 450	10
Diclofenac	PPCPs	Low Concentration	100	ND – 305	11
Digoxigenin	PPCPs	< 10 % Detects	NA	ND	2
Digoxin	PPCPs	< 10 % Detects	NA	ND	2
Diltiazem	PPCPs	Low Concentration	37	ND – 225	2
Diphenyltin	Organotins	Low Concentration	NA	ND – 400	12
Enrofloxacin	PPCPs	Low Concentration	27	ND – 66	2
Erythromycin	PPCPs	Low Concentration	36	ND – 180	2
Ethanol,2-butoxy-,phosphate	Plasticizers	Low Concentration	416	ND – 650	1
Famotidine	PPCPs	Low Concentration	20	ND – 75	11
Flumequine	PPCPs	< 10 % Detects	NA	ND	2
Fluoxetine	PPCPs	Low Concentration	245	ND – 3,140	1, 2
Gemfibrozil	PPCPs	Low Concentration	210	ND – 2,650	1, 2
Glibenclamide	PPCPs	Low Concentration	60	ND – 150	11
Hexchlorophene (HCP)	Phenols	Low Concentration	394	23 – 1,190	13
Hydrochlorothiazide	PPCPs	Low Concentration	10	ND – 15	11
Hydroxybiphenyls	Phenols	Low Concentration	NA	ND – 172	14
Isochlortetracycline	PPCPs	< 10 % Detects	81	ND – 3,140	2
Ketoprofen	PPCPs	Low Concentration	13	ND – 55	11
Lincomycin	PPCPs	< 10 % Detects	29	ND – 33	2
Lomefloxacin	PPCPs	< 10 % Detects	23	ND – 40	2

NA = not available; ND = not detected

Data Sources:

1 (Kinney et al., 2006)
2 (U.S. EPA, 2009i)
3 (Barnabe et al., 2008)
4 (Herren and Berset, 2000)
5 (Gielen, 2007)
6 (Jones-Lepp and Stevens, 2007)
7 (Kaleta et al., 2006)
8 (Petrovic and Barcelo, 2000)
9 (Kupper et al., 2004)
10 (Torslov et al., 1997)
11 (Radjenovic et al., 2009)
12 (Fent, 1989)
13 (Lee and Peart, 2002)
14 (Bolz et al., 2001)
15 (Smyth et al., 2007)
16 (Berset et al., 2000)
17 (Mumma et al., 1984)
18 (Spongberg and Witter, 2008)
19 (Xu et al., 2007)

Table 2-3. Trace Organic Chemicals Excluded from Data Gap Analysis Based on Occurrence Data (continued).

Chemical	Chemical Class	Reason for Exclusion	Max Mean or Median (µg/kg)	Concentration Range (µg/kg)	Data Sources
Loratidine	PPCPs	Low Concentration	78	ND – 195	2
Mefenamic acid	PPCPs	Low Concentration	15	ND – 40	2
Metformin	PPCPs	< 10 % Detects	542	ND – 1,160	2
Methamphetamine	PPCPs	Low Concentration	NA	ND	6
Monophenyltin	Organotins	Low Concentration	NA	ND – 100	12
Musk Ambrette (MA)	PPCPs	Low Concentration	19	ND – 31	15
Musk Moskene (MM)	PPCPs	Low Concentration	NA	ND – 6	15
Musk Tibetene (MT)	PPCPs	Low Concentration	NA	ND – 67	15
Musk Xylene (MX)	PPCPs	Low Concentration	33	NA	16
N-nitrosopyrrolidine	Nitrosamines	Low Concentration	NA	ND – 4	17
Naproxen	PPCPs	Low Concentration	82	ND – 1,020	2
Norfloxacin	PPCPs	Low Concentration	250	ND – 1,290	2
Norgestimate	PPCPs	< 10 % Detects	NA	ND	2
Ormetoprim	PPCPs	< 10 % Detects	NA	ND – 6	2
Oxacillin	PPCPs	< 10 % Detects	NA	ND	2
Oxolinic acid	PPCPs	< 10 % Detects	5	ND – 39	2
Oxytetracycline	PPCPs	Low Concentration	57	ND – 467	2
Paroxetine	PPCPs	Low Concentration	20	ND – 50	11
Penicillin G	PPCPs	< 10 % Detects	NA	ND	2
Penicillin V	PPCPs	< 10 % Detects	NA	ND	2
Propranolol	PPCPs	Low Concentration	15	ND – 35	11
Ranitidine	PPCPs	Low Concentration	51	ND – 2,250	2
Roxithromycin	PPCPs	< 10 % Detects	8	ND – 23	2
Salicylic Acid	PPCPs	Low Concentration	175	ND – 253	18
Sarafloxacin	PPCPs	< 10 % Detects	266	ND – 1,980	2
Sulfachloropyridazine	PPCPs	< 10 % Detects	12	ND – 59	2
Sulfadiazine	PPCPs	< 10 % Detects	13	ND – 140	2
Sulfadimethoxine	PPCPs	< 10 % Detects	3	ND – 62	2
Sulfadimidine	PPCPs	Low Concentration	NA	ND – 31	19
Sulfamerazine	PPCPs	< 10 % Detects	5	ND – 6	2
Sulfamethazine	PPCPs	< 10 % Detects	7	ND – 23	2
Sulfamethizole	PPCPs	< 10 % Detects	NA	ND	2
Sulfamethoxazole	PPCPs	Low Concentration	19	ND – 651	2
Sulfathiazole	PPCPs	< 10 % Detects	11	ND – 21	2
Sulfisoxazole	PPCPs	Low Concentration	16	ND – 22	18
Thioridazine	PPCPs	Low Concentration	NA	ND	7
Trimethoprim	PPCPs	Low Concentration	4	ND – 204	1, 2
Tri-*n*-butylphosphate	Phosphate Esters	Low Concentration	252	ND – 2,400	10
Triphenylphosphate	Phosphate Esters	Low Concentration	284	ND – 1,900	10
Triphenyltin	Organotins	Low Concentration	NA	ND – 300	12
Tylosin	PPCPs	< 10 % Detects	NA	ND	2
Virginiamycin	PPCPs	Low Concentration	133	ND – 469	2
Warfarin	PPCPs	< 10 % Detects	33	ND – 92	1

NA = not available; ND = not detected

Data Sources:

1 (Kinney et al., 2006)
2 (U.S. EPA, 2009i)
3 (Barnabe et al., 2008)
4 (Herren and Berset, 2000)
5 (Gielen, 2007)
6 (Jones-Lepp and Stevens, 2007)
7 (Kaleta et al., 2006)
8 (Petrovic and Barcelo, 2000)
9 (Kupper et al., 2004)
10 (Torslov et al., 1997)
11 (Radjenovic et al., 2009)
12 (Fent, 1989)
13 (Lee and Peart, 2002)
14 (Bolz et al., 2001)
15 (Smyth et al., 2007)
16 (Berset et al., 2000)
17 (Mumma et al., 1984)
18 (Spongberg and Witter, 2008)
19 (Xu et al., 2007)

2.2.5 Inclusions Based on Risk Perception and Known Data Gaps

A principal driver for this analysis was that much is still unknown regarding specific TOrCs that may have adverse effects on humans or the environment. Three specific classes of TOrCs were identified that have received considerable scientific and public attention in recent years and for which data gaps regarding their presence and impacts exist: brominated flame retardants (BFRs), perfluorochemicals (PFCs) and their precursors, and synthetic steroidal or endocrine disrupting chemicals.

The high concentrations of BFRs in consumer products (Watanabe and Sakai, 2003) and apparent accumulation in wildlife (Law et al., 2003) have raised concerns about BFRs in biosolids. Given the scientific and public interest, the entire class of BFRs was included, even though specific BFRs have low occurrence values. Many of the studies on polybrominated diphenyl ethers (PBDEs; a subclass of BFRs) have been conducted on a small suite of congeners. Thus, PBDE congeners were the focus of this review.

The presence of relatively low concentrations of PFCs in sewage sludge (Higgins et al., 2005) suggest minimal risk from land application of biosolids. However, recent reports of PFC contamination of ground and surface waters (Renner, 2009; U.S. EPA, 2009g) following land-application of biosolids raises concerns. Importantly, the biosolids applied to these soils were likely contaminated with high levels of PFCs from industrial sources and, thus, were possibly unique. In addition, scientific debate is ongoing regarding the ability of PFC precursors, such as perfluoroalkyl-based polymers, to degrade in the environment to form PFCs (Washington et al., 2009). Some of the precursors have been detected in sludge (D'Eon et al., 2009), while other PFC precursors may also be present in sewage sludge, but have simply not been detected due to analytical limitations. The potential presence of PFC precursors in sewage sludge and biosolids suggests a data gap analysis with respect to PFCs in biosolids is warranted.

Finally, the widespread documentation of the endocrine-disrupting effects of many steroidal chemicals (Hanselman et al., 2003) suggests that even low levels of these TOrCs in biosolids may result in adverse environmental impacts. Because of its endocrine disrupting potential, the synthetic steroidal chemicals 17α-Ethinyl estradiol (EE2) and mestranol (MeEE2) were included as high priority TOrCs, whereas many of the naturally occurring steroidal chemicals were listed as low priority TOrCs.

2.3 Prioritization

Time and resource constraints necessitated prioritization of TOrCs for detailed analysis of data availability and knowledge gaps. Figure 2-1 provides a schematic describing the selection and prioritization of TOrCs for this study. With the exception of the naturally-occurring steroidal chemicals, any TOrC characterized by known data gaps or significant risk perception was included as a high priority TOrC. Included as low priority TOrCs were chemicals with maximum mean or median concentrations less than 1000 μg/kg, and/or whose range of observed concentrations exceeded 1000 μg/kg in either of the two national surveys.

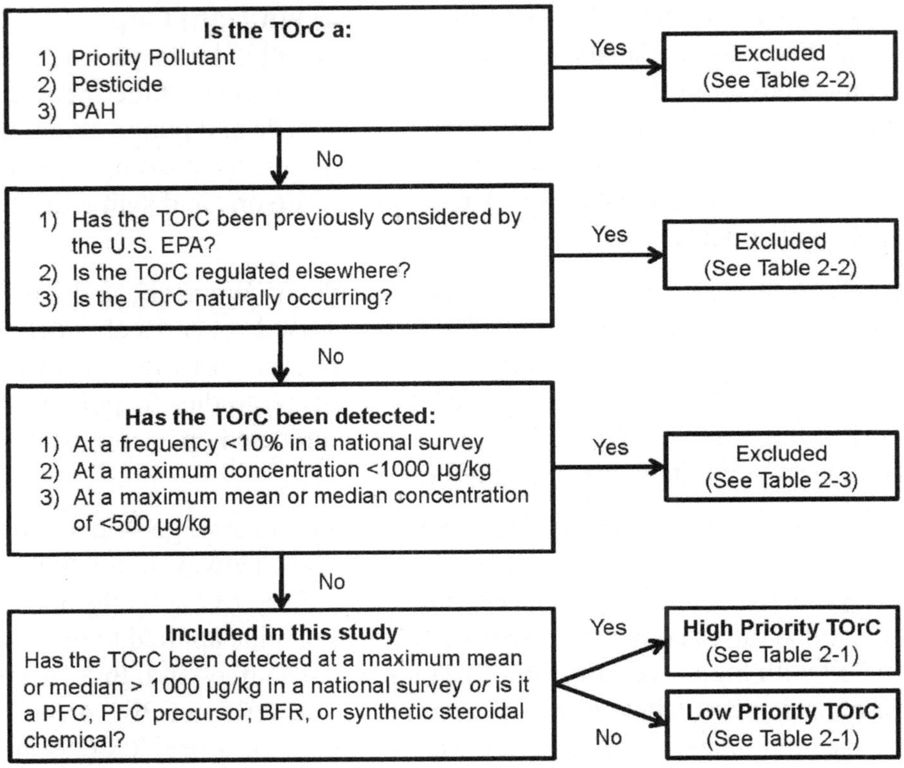

Figure 2-1. Schematic for Selection and Prioritization of TOrCs for this Study.

CHAPTER 3.0

ASSESSING THE RISK OF BIOSOLIDS-BORNE TRACE ORGANIC CHEMICALS IN SOILS

3.1 Introduction

The U.S. EPA considers risk to be the chance of harmful effects to human health or ecological systems resulting from exposure to an environmental stressor (e.g., any physical, chemical, or biological entity that can induce an adverse response). Risk assessments are used to quantify the magnitude of these risks to human and ecological receptors. Ideally, risk assessments would be based on strong scientific knowledge of exposure and inherent toxicity of a chemical. In reality, though, information is usually limited for key aspects of the risk assessment. Due to limited data, risk assessors must make estimates and use professional judgment when performing risk calculations. Therefore, a risk assessment should be transparent of the uncertainties in the calculations. Information from a risk assessment is then used in the risk management process, which evaluates how best to protect human health and ecological systems. Risk management leads to the action taken on potential human and ecological risk. Various factors feed into the risk management process (e.g., scientific, economic, legal, social, technological and political). Risk assessments are only one of many considerations that go into a risk management decision.

Many schemes have evolved over the years for assessing the risk of biosolids-borne trace organic chemicals (TOrCs) in soil environments, but the simplest way to assess risk is to compare the environmental exposure concentration to an acceptable environmental effects concentration. If the environmental exposure concentration is less than the acceptable effects concentration, then minimal risk is expected. To aid in the assessment process, multimedia models are used to evaluate the fate, exposure, and potential effects of TOrCs following biosolids amendment to soils. These models are used to estimate TOrC exposure in the matrix of interest, define safe levels of exposure based on known or predicted toxicological effects, and provide an assessment of the potential risk from the predicted exposure and effects concentrations.

What follows is a brief description of risk assessment procedures used in the United States and Europe for predicting exposure and effects of TOrCs following the application of biosolids to soil. While the focus of this review is on assessing TOrC risk in the United States, procedures from Europe are included to ensure that the best techniques are being considered when assessing chemical risk. Differences in the approaches are highlighted, data requirements are identified, and enhanced risk modeling needs are recommended.

3.2 U.S. EPA - Part 503

The U.S. EPA developed a risk-based approach for assessing potential impacts of biosolids-borne chemicals to human and ecological health (40 CFR Part 303; U.S. EPA, 1993a). The approach is based on procedures described by the National Academy of Science (NAS, 1983) and includes four basic steps: 1) hazard identification (toxicity assessment); 2) exposure assessment; 3) dose-response evaluation (human, plant and animal); and 4) risk characterization (Risk = Hazard x Exposure). Initially, the U.S. EPA used a deterministic risk assessment approach to evaluate potential impacts of biosolids-borne chemicals. The deterministic risk assessment used discrete, single-point input values for emissions, bioavailability factors, uptake slopes, dose-response relationships, characteristics of the target population, and variables to calculate risk for a highly exposed individual (U.S. EPA, 1995). The assessment approach was later refined to include a probabilistic risk assessment methodology. The probabilistic risk assessment uses a distribution of input parameters in the mathematical simulation models (U.S. EPA, 2003b). A mixture of average and upper bound assumptions are used to identify reasonable maximum exposure receptors (e.g., human, plants, or animals). A Monte Carlo analysis is used to quantify uncertainty in the risk calculation. The main advantage of a probability risk assessment is that the degree of conservatism can be more accurately determined. The major disadvantage of a probabilistic risk assessment is that it requires significantly more data to execute. With either approach, deterministic or probabilistic risk assessment, a policy decision needs to be made with regard to the level of acceptable risk. Conservative assumptions are made to ensure adequate protection to human health and environmental quality.

The risk characterization processes for Part 503 describe 14 exposure pathways for assessing the risk of biosolids amendment to soil (Table 3-1). Human exposure pathways include consumption of vegetables grown in biosolids amended soils, consumption of meat and dairy products fed with vegetation grown in biosolids amended soil, runoff and infiltration to drinking water sources, and direct ingestion of amended soil. Ingestion of the amended soil has sometimes been the limiting pathway for some contaminants, while other pathways (e.g., groundwater as a source of drinking water) have been identified for other contaminants. Ecological exposure pathways include direct exposure to soil-dwelling organisms and food crops, as well as exposure via the food web for mammals and birds. The exposure pathways are conceptually depicted in Figure 3-1.

Table 3-1. Exposure Pathways Described in Part 503 Regulations.

Pathways to the Environment
Sludge-soil-soil biota
Sludge -soil-plant
Sludge-soil-soil biota-predator
Pathway to Livestock and Human
Sludge-soil-child
Sludge-soil-animal
Sludge-soil-plant-animal
Sludge-soil-plant-human
Sludge-soil-animal-human
Sludge-soil-plant-animal-human
Pathways to Human
Sludge-soil-airborne particle-human
Sludge-soil-atmosphere-human
Sludge-soil-surface runoff-surface water-human
Sludge-soil vadose zone-groundwater-human

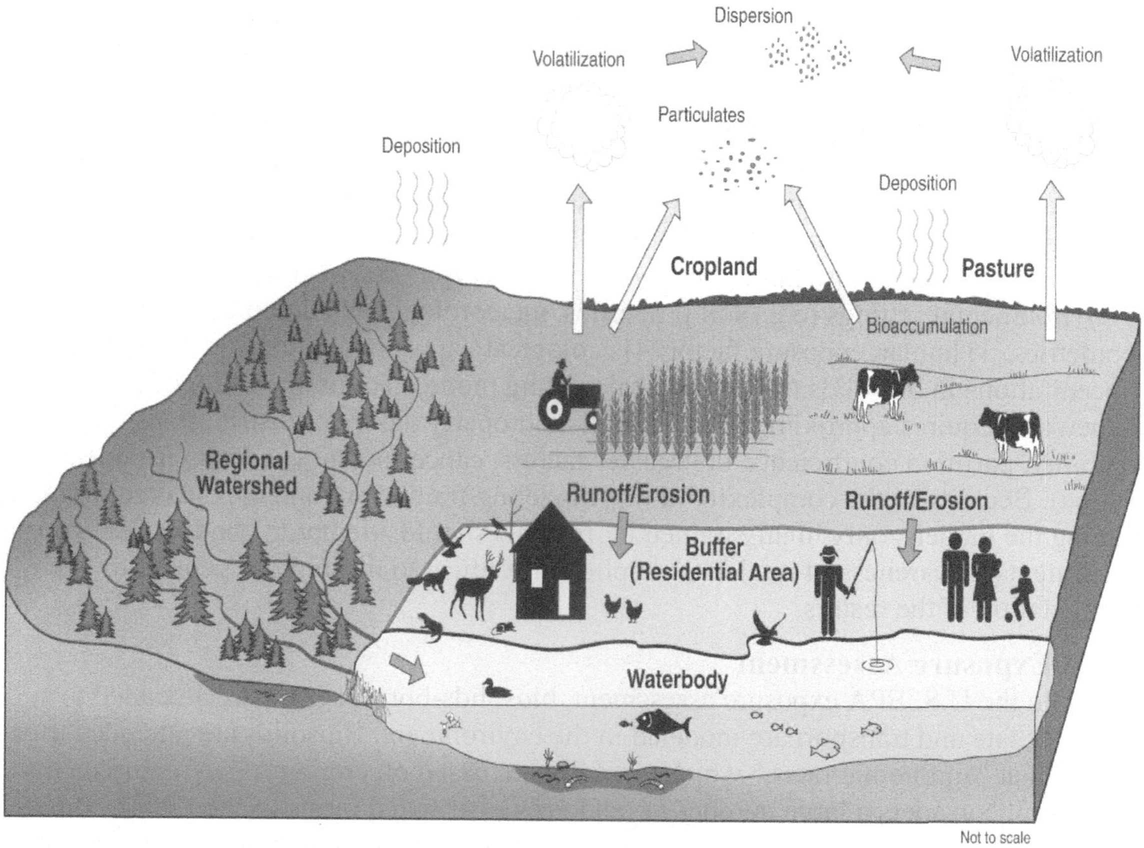

Figure 3-1. Conceptual Diagram of the Exposure Pathways Considered in the U.S. EPA Risk Assessment.

An overview of the U.S. EPA risk assessment methodology, including algorithms and associated input parameters, is provided in "Technical Background Document for the Sewage Sludge Exposure and Hazard Assessment" (RTI, 2008). Briefly, the methodology for assessing biosolids risk uses a series of algorithms that mathematically represent each exposure pathway. Each algorithm contains numerous parameters related to a chemical's fate, transport, exposure, and potential effects to receptors. Information for some of the parameters may be available, while others may need to be estimated using various methodologies (empirical or *in silico*).

The U.S. EPA approach uses a multimedia, multipathway, multireceptor risk modeling framework to characterize potential chemical hazards associated with the land application of biosolids. The framework includes nine modules that require chemical-specific property data:

- Source module
- Groundwater module
- Watershed module
- Surface water module
- Aquatic food web module
- Terrestrial food web module
- Farm food chain module
- Human risk module
- Ecological risk module

The framework implements a Monte Carlo simulation approach that generate a distribution of risk for chemicals in which concentration data are available or it generates a distribution of allowable biosolids concentrations based on U.S. EPA's target risk levels. Thus, the framework can be implemented in either a "forward-calculating" mode (i.e., estimate risk based on the concentration distribution in biosolids) or a "back-calculating" mode (i.e., risk-based concentrations are estimated in the absence of concentration data). In both cases, the model is typically implemented for a national-scale assessment of the terrestrial environment and two aquatic scenarios (farm pond and index reservoir) that accounts for the variability in: 1) environmental settings (e.g., soil properties, meteorology, and depth to groundwater); 2) field geometries; 3) human exposure factor; 4) ecological exposure factors; and 5) chemical concentrations in biosolids (in forward-calculating mode). To run the full model simulation, the framework requires approximately 60 chemical property values or distributions (e.g., water solubility, partition coefficients, biotransfer factors, cancer potency, and environmental quality criteria). Because of the complexity of this modeling framework and the data requirements for running the models, only highly trained individuals should attempt to conduct the exposure assessments. Awareness of model assumptions is critical to meaningful generation and interpretation of the results.

3.2.1 Exposure Assessment

In the U.S. EPA exposure assessment, biosolids-borne TOrCs are amended to soil and then their fate and transport are modeled in the environment. Biosolids are amended to pasture or crop land at "agronomic rates," which typically are based on crop nitrogen requirements. The assessment considers a large number of randomly distributed farms located in 41 climate regions across the United States. The multiple climate zones are used to address inherent variability of soil properties, meteorological conditions, and crop types. These scenarios also include transport of TOrCs via runoff into two bodies of water: 1) an index reservoir used for drinking water; and 2) a farm pond with ecological receptors. The family is assumed to live on the farm continuously for 70 years and to consume only food raised from the biosolids-amended fields.

The assessment approach assumes that the TOrC concentration in the biosolids is known (forward-calculating mode). The biosolids concentration is usually derived by analyzing a variety of biosolids from across the U.S. for contaminants of interest and then using the 95th percentile concentration as the biosolids concentration being amended to the soil environment (U.S. EPA, 2009f; U.S. EPA, 2009i). If measured biosolids contaminant concentrations are not available, then risk-based concentrations are estimated in the absence of concentration data or source modeling can be used to estimate their concentration in biosolids. Several wastewater treatment plant models are available to assist in predicting TOrC concentrations in biosolids. These models include ASTreat (McAvoy et al., 1999), SimpleTreat (Struijs, 1996) and ToxChem (Hydromantis, 2009a). Information on the chemical fate processes of sorption, biodegradation, and volatilization during wastewater treatment are needed to run these simulation models. While these models have been sufficiently validated for predicting effluent TOrC concentrations, their use for predicting biosolids concentrations is still somewhat limited.

Once the biosolids TOrC concentrations are known or predicted, fate and transport modeling is used to assess their movement in soil following release to agricultural fields. Soil erosion and runoff to the farm pond, as well as leaching to groundwater for assessing drinking water concentrations are simulated. The TOrC leachate concentrations are adjusted to account for dilution by ambient groundwater using a protective dilution-attenuation factor. Movement of the contaminant to, and in, groundwater is influence by soil and aquifer conditions, which are taken into account by simulating the 41 climate zones. Site specific data for meteorological conditions and soil type are also needed for the 41 climate zones.

Chemical-specific data needed to run the exposure assessment models are presented in Table 3-2. These data are critical for assessing the fate, transport, and exposure of TOrCs following biosolids amendment to soil. More specifically, these parameters are used to calculate sorption, volatilization, and degradation in the soil and aquatic environments. More detail on the use of these parameters is provided in "Technical Background Document for the Sewage Sludge Exposure and Hazard Assessment" (RTI, 2008).

Table 3-2. Chemical-Specific Data Needed for Exposure Assessment Modeling.

Parameter Abbreviation	Physical-Chemical Properties
Density	Density
Da	Diffusivity in air
Dw	Diffusivity in water
HLC	Henry's law constant
$k_{aer\text{-}soil}$	Aerobic biodegradation rate in soil
$k_{aer\text{-}sw}$	Aerobic biodegradation rate in surface water
$k_{anaer\text{-}sed}$	Anaerobic biodegradation rate in sediment
$k_{anaer\text{-}sed}$	Anaerobic biodegradation rate in sediment
K_d	Sorption distribution coefficient
k_h	Hydrolysis rate
K_{oc}	Organic carbon-water distribution coefficient
K_{ow}	Octanol-water distribution coefficient
MP	Melting point
MW	Molecular weight
P_c	Critical pressure
pKa	Dissociation constant, acid
Sol	Water solubility
T_b	Boiling point
T_c	Critical temperature
VP	Vapor pressure

3.2.2 Hazard Assessment

Ecological receptors in the hazard assessment for land application of biosolids include both terrestrial species (from direct exposure following soil amendment) and aquatic species (from runoff to a farm pond). The ecological receptors considered to be most importance are: 1) fish, aquatic invertebrates, aquatic plants, amphibians, and sediment biota in the farm pond; 2) soil invertebrates and plants in the agricultural field; and 3) mammals and birds in contact with the agricultural field and farm pond. Suggested species for consideration in the hazard assessment are provided in "Technical Background Document for the Sewage Sludge Exposure and Hazard Assessment" (U.S. EPA, 2003b). Because of the large number of ecological species that could potentially be exposed following biosolids amendment, the hazard assessment typically focuses on a smaller number of indicator organisms thought to represent the most exposed or most sensitive species from three trophic levels (U.S. EPA, 1998a). A select list of species for mammal and bird wildlife is presented in Table 3-3.

Table 3-3. Selected List of Ecological Receptors – Mammal and Bird Wildlife Species.

Species	Feeding Guild[a]	Trophic Level[b]
American robin	O	T2
American woodcock	O	T2
Canada goose	H	T1
Coyote	O	T3
Eastern cotton tail rabbit	H	T1
Meadow vole	H	T1
Mink	C	T2
Raccoon	O	T2
Red-tailed hawk	C	T3
Tree swallow	O	T2
White-tailed deer	H	T1

[a] Feeding guild: C = carnivore, H = herbivore, O = omnivore
[b] Trophic level: T1 = prey, not predator; T2 = both predator and prey; T3 = top predator, not prey

The U.S. EPA's Office of Prevention, Pesticides and Toxic Substances (OPPTS) developed test guidelines for assessing ecological effects (see Table 3-4 for a select list of OPPTS test methods). The most relevant endpoints for conducting an effects assessment are reproduction, growth, and development. However, for most chemicals only mortality (e.g., 50% lethal dose [LD_{50}] or 50% lethal concentration [LC_{50}]) data are available. These data are generally not considered sufficient to protect ecosystem health, but in the absence of other data (e.g., chronic toxicity or reproduction effects) provide a basis for conducting screening level assessments and for setting ecological benchmarks when appropriate safety factors are applied.

Table 3-4. Selected List of Acceptable Test Methods for Assessing Ecological Effects.

Test Description	Test Method
Terrestrial Vertebrates	
Avian acute oral toxicity	OPPTS 850.2100 (OECD 223)
Avian short-term dietary toxicity	OPPTS 850.2200 (OECD 205)
Avian reproduction	OPPTS 850.2300 (OECD 206)
Terrestrial Invertebrates	
Honeybees toxicity of residues on foliage	OPPTS 850.3030
Honeybees acute contact toxicity	OPPTS 850.3020 (OECD 214)
Soil Macroorganisms	
Earthworm field	ISO 11268
Soil Microorganisms	
Soil microbial community toxicity	OPPTS 850.5100
Terrestrial Plants	
Tier 1 – seedling emergence	OPPTS 850.4100
Tier 1 – vegetative vigor	OPPTS 850.4150 (OECD 227)
Tier 2 – seedling emergence	OPPTS 850.4225
Tier 2 – vegetative vigor	OPPTS 850.4250
Tier 3 – field study	OPPTS 850.4300
Aquatic Fauna	
Daphnid acute toxicity	OPPTS 850.1010
Fish acute toxicity	OPPTS 850..1075 (OECD 203)
Daphnid chronic toxicity	OPPTS 850.1300 (OECD 202)
Fish early life stage toxicity	OPPTS 850.1400 (OECD 210)
Fish BCF	OPPTS 850.1730 (OECD 305)
Whole sediment acute toxicity	OPPTS 850.1735
Aquatic food chain transfer	OPPTS 850.1850

For human receptors, the effects assessment assumes that each family member consumes only farm-raised food from agricultural land amended with biosolids and fish from the nearby farm pond. This assumption allows for a conservative assessment. Other human exposure pathways considered are drinking water from an onsite groundwater well, inhalation of ambient air and dust, and inhalation of shower indoor air (groundwater source). The human health assessment also considers bioaccumulation and biomagnification of TOrCs through food webs.

The U.S. EPA's OPPTS also developed test guidelines for assessing human health effects (see Table 3-5 for a select list of OPPTS test methods). The most relevant endpoints for conducting a human effects assessment are acute toxicity, subchronic toxicity, chronic toxicity, genetic toxicity, and neurotoxicity.

Table 3-5. Selected List of Acceptable Test Methods for Assessing Human Health Effects.

Test Description	Test Method
Acute Toxicity	
Acute oral toxicity	OPPTS 870.1100 (OECD 401)
Acute dermal toxicity	OPPTS 870.1200 (OECD 402)
Subchronic Toxicity	
90-day oral toxicity in rodents	OPPTS 870.3100 (OECD 408)
90-day dermal toxicity	OPPTS 870.3250 (OECD 411)
Chronic Toxicity	
Chronic toxicity	OPPTS 870.4100 (OECD 452)
Carcinogenicity	OPPTS 870.4200 (OECD 451)
Genetic Toxicity	
Bacteria reverse mutation	OPPTS 870.5100 (OECD 471)
In-vitro mammalian cell gene mutation	OPPTS 870.5300 (OECD 476)
Rodent dominant lethal assay	OPPTS 870.5450 (OECD 478)
Neurotoxicity	
Neurotoxicity screening battery	OPPTS 870.6200 (OECD 424)
Developmental neurotoxicity study	OPPTS 870.6300
Special Studies	
Metabolism and pharmacokinetics	OPPTS 870.7485 (OECD 417)

3.2.3 Risk Characterization

Risk characterization involves assessing relationships among the release of a chemical, and its exposure and potential effects to a receptor. The overall analysis includes a rationale for the methods and models used, identification of data gaps, and quantification of uncertainties associated with the data used in the analysis. The uncertainty analysis determines an estimate of uncertainty (standard deviation) in the expected concentration of the output variables (e.g., mean exposure concentration) due to uncertainty in model parameters, inputs, and initial state. Data variability and uncertainty are accounted for by applying safety factors in the calculations. Initially, a screening level assessment is conducted. Depending on the degree of uncertainty in the screening assessment, higher level assessments may be needed (e.g., refinement of model inputs or the use of measured field data).

Multimedia fate and transport models are used to estimate contaminant exposure and inputs to the food web transfer models. The biotransfer models include predicting the uptake of TOrCs by plants grown in biosolids-amended fields, accumulation by fruits and vegetables, and uptake by beef and dairy cattle that consume forage and silage grown on the biosolids-amended fields. Contaminants in the biosolids may also erode and runoff into the farm pond where TOrCs could accumulate in fish.

The human health assessment assumes that family members consume beef and dairy products from cattle that forage on amended pasture land and consume silage raised on the farm.

Other exposure pathways include ingestion of produce raised on the farm, fish caught in the nearby pond, soil from the amended fields (unwashed food crops or pica behavior), and drinking water from onsite groundwater wells. Biotransfer factors (BTFs) are needed for all of these exposure pathways. A list of the biotransfer factors needed in the human health assessment is presented in Table 3-6.

Table 3-6. Biotransfer Factors Needed in the Human Health Assessment.

Parameter Abbreviation	Bioconcentration and Biotransfer Factors
BCF_beef	Bioconcentration factor in beef
BCF_eggs	Bioconcentration factor in eggs
BCF_fish	Bioconcentration factor in fish
BCF_milk	Bioconcentration factor in milk
BCF_pork	Bioconcentration factor in pork
BCF_poultry	Bioconcentration factor in poultry
BrExfruit	Plant-soil bioconcentration factor in exposed fruit
BrExveg	Plant-soil bioconcentration factor in exposed vegetables
BrForage	Plant-soil bioconcentration factor in forage
BrGrain	Plant-soil bioconcentration factor in grain
BrProfruit	Plant-soil bioconcentration factor in protected fruit
BrProveg	Plant-soil bioconcentration factor in protected vegetables
BrRoot	Plant-soil bioconcentration factor for roots
BrSilage	Plant-soil bioconcentration factor for silage
Bs	Bioavailability of contaminant on the soil relative to vegetation
RCF	Root concentration factor

The ecological assessment is more complex than the human health assessment due to the larger number of receptors, food web linkages (predatory prey links), and function within the ecosystem. The ecological receptors of greatest interest in the terrestrial environment are soil bacteria, plants, invertebrates, birds, and mammals. Since ecological risk assessments cannot evaluate all of the exposed species, the assessments typically focus on a smaller number of indicator organisms that are representative of the most exposed or most sensitive species (U.S. EPA, 1998a). The most relevant ecological effects endpoints for the indicator species are growth, survival, and reproduction.

Biological-uptake of TOrCs is also considered in the ecological assessment. This assessment assumes that all of a receptor's diet comes from either the farm pond or the agricultural field. The exposure dose is calculated as a function of ingestion rate, body weight, and concentrations in the various food sources. Ingestion rates are needed for water consumption, soil ingestion, vegetation (fruits, forage, gain, roots) ingestion, and prey (birds, mammals, invertebrates, fish) transfers. Body weights and ingestion rates are available from the "Wildlife Exposure Factors Handbook" (U.S. EPA, 1993b).

Terrestrial animal bioaccumulation factors (BAFs) are generally not available and suitable *in silico* relationships have not been established. When data is limited, small mammal BAFs are used for all terrestrial vertebrate prey and earthworm BAFs are used for all terrestrial invertebrate prey. In the absence of measured data, a BAF of 1 is assumed for all terrestrial vertebrate and invertebrate prey. For some chemicals, this assumption may not be conservative due to biomagnification at higher trophic levels. Terrestrial plant and animal BAFs can also be estimated using the Biosolids Amended Soil Level IV model (Trent University, 2009). Since this model has not been fully evaluated for the TOrCs identified in this study, caution should be taken when using it for predicting terrestrial BAF values.

For the aquatic food web, primary producers (e.g., algae) are grazed by zooplankton and the zooplankton are consumed by planktivorous fish. The planktivorous fish are then consumed by carnivorous fish. Aquatic plants and animals can also be consumed by terrestrial herbivores, omnivores, and carnivores.

3.2.4 Data Needs

Key parameters needed for conducting U.S. EPA's biosolids risk assessment and their relative importance is presented in Table 3-7. The methodology requires information on chemical properties, volatilization, degradation rates, dissociation constants (pK_a), organic carbon normalized soil-water partition coefficients (K_{oc}), soil-water partition coefficients (K_d), bioconcentration (BCFs) and BAFs for ecological assessments, and biotransfer factors for human health assessments. If measured values are not available, then estimates are made using accepted *in silico* methods. Plant biotransfer factors can be derived from the "Methodology for Assessing Health Risks Associated with Multiple Pathways of Exposure to Combustor Emissions" (U.S. EPA, 1998b). Beef and milk biotransfer factors are available in "Methodology for Predicting Cattle Biotransfer Factors" (RTI, 2005). Aquatic food chain (e.g., fish) bioconcentration factors can be estimated using EPIWIN (U.S. EPA, 2008a).

The methods for predicting biological-uptake require a log octanol-water partitioning coefficient (K_{ow}) value and utilize relationships between K_{ow} and bioaccumulation developed for hydrophobic compounds. Since not all TOrCs are hydrophobic compounds, the current models may not be applicable for some of the TOrCs identified in this review. Similarly, the various relationships developed for predicting K_{oc} and K_d from K_{ow} are highly dependent on chemical class (structural similarities) and should be judiciously used. Volatilization via Henry's Law constant can be estimated from chemical properties using EPIWIN (U.S. EPA, 2008a). Water solubility and vapor pressure are also used to predict the Henry's Law constant. Degradation rates (biodegradation, hydrolysis, and photolysis) are difficult to predict and thus are best measured. When no empirical data are available, a default value of zero (no degradation) is assumed. This assumption makes the assessment very conservative.

The minimum data set required to run U.S. EPA's risk model is presented in Table 3-8. A more rigorous sensitivity analysis is needed to quantify the effect of changes in model parameters on the model outcome. For the minimum data set parameters, values should only be used if: 1) they are produced using accepted analytical techniques, published in peer reviewed studies, or reports; or 2) they can be estimated using U.S. EPA-approved or other peer reviewed methods. The minimum data set could be refined by establishing a screening model methodology that focuses on a specific subset of exposure pathways considered relevant to the type of chemical being evaluated (e.g., bioaccumulative chemicals). However, such a methodology does not currently exist and, therefore, the minimum data set focuses on parameters currently required to run the risk assessment model. The table also includes references to *in silico* models (e.g., EPI Suite, SPARC) that could be used to estimate model parameters. Further investigation is needed, though, to determine whether these estimation techniques could be used for the TOrCs identified in this study. Direct measurement of these parameters is preferred, thus acceptable test methods (e.g., OPPTS, OECD) are also included in the table.

3.3 European Union Technical Guidance Document

In Europe, the use of biosolids on agricultural land is regulated under Directive 86/278/EEC. New and existing chemicals are assessed for possible concern using risk assessments described in the European Union (EU) Technical Guidance Document (TGD; CEC, 1996). The methodology for assessing risk compares the predicted exposure concentration (PEC) to the predicted no effect concentration (PNEC) in the environmental compartment of interest. If the PEC to PNEC ratio is greater than one (PEC/PNEC >1), then there is concern for environmental effects. Either the risk assessment can be refined by improving the input data (using a tiered testing strategy that reduces uncertainty in the prediction), or risk reduction measures could be implemented. The risk assessment described in the EU TGD is generic in the sense that it assesses exposure and effects on a typical local or regional scale using a standard European scenario.

Table 3-7. Key Parameters Needed in the U.S. EPA Risk Assessment.

Module	Key Parameters	Relative Importance
Human risk	Health benchmarks for cancer and noncancerous endpoints (e.g., cancer potency factor, noncancerous reference dose)	Absolute requirement to estimate health risk; can use most toxic chemical from class to represent all chemicals in a class. Model results are linear with respect to benchmarks.
Ecological risk	Ecological benchmarks for relevant endpoints for populations and communities, includes aquatic and terrestrial organisms	Full model includes multiple trophic levels; however, can focus on high contact receptor groups (e.g., benthos) and most highly exposed species (e.g., shrew). Model results are linear with respect to benchmarks.
Source (field)	Partition coefficient, water solubility, dissociation constant, diffusivity, degradation rate	Some properties can be estimated from others (e.g., Henry's Law constant); however, certain basic properties are required. Some properties (e.g., biodegradation) can be set at conservative default values based on science-policy and/or first principles. Risk estimates are moderately to highly sensitive to chemical property values in the source module.
Surface water	Similar properties are required as those needed by the source (field) module; however, the dynamic waterbody module is sensitive to degradation rates (e.g., hydrolysis, photolysis)	As with the source module, some properties can be estimated from others, and default values based on science-policy can be used to address data gaps. For some organic chemicals, the ecological risk results are moderatively sensitive to chemical property values, especially degradation rates in water.
Groundwater	Similar properties to the surface water module are needed, but diffusion coefficient is particularly important	Chemical property requirements are less for the groundwater module, and most can be estimated from basic properties. The groundwater module is particularly sensitive to parameters on degradation rates and on diffusion potential. Risk results are moderately sensitive to property values for the groundwater module.
Watershed	Few chemical-specific parameters are required by the watershed module, but the degradation rates in soil are important in estimating soil concentrations in the buffer zone adjacent to the field	Data on chemical degradation rates (as with many degradation rates) in soil systems are often not available, especially for TOrCs that are not well characterized. Conservative assumptions can be used to estimate degradation, but the risk estimates are not very sensitive to changes in parameter values for this module.
Aquatic Food Web	The aquatic food web module requires uptake and accumulation values such as bioconcentration and bioaccumulation.	For most TOrCs the BCFs, BAFs, and BMFs can be estimated using regression equations or food web models based on the octanol-water partition coefficient. Ionizable chemicals require special treatment (e.g., BCF is strongly dependent on speciation for weak bases).
Farm Food Chain	The farm food chain module requires biotransfer factors (BTFs) for plants, beef, dairy, etc	For most TOrCs the BTFs can be estimated using published estimation techniques and basic properties (e.g., molecular weight, K_{ow}). Estimation methods are not as well established for exposure pathways involving pork, chicken, and eggs. However, the risk estimates tend to be weakly sensitive to biotransfer factors for the majority of chemical pollutants (exceptions include dioxins, PCBs, and certain other persistent organic compounds).
Terrestrial Food Web	The terrestrial food web requires bioaccumulation factors (e.g., soil-to-worm) and biotransfer factors (for plants to estimate bioaccumulation) to estimate the concentrations in various plants, livestock, and prey	For most TOrCs the BAFs and BTFs can be estimated using the same methods as for the Farm Food Chain. To estimate exposure concentrations in various prey species, few estimation methods have been developed beyond earthworms, and a default value of 1 is often assumed where no data are available for screening or prioritization purposes.

Table 3-8. Minimum Data Set Required for the U.S. EPA Risk Assessment.

Parameter	Module(s)	Test Methods	Estimation Techniques
Health benchmark	Human risk	Cancer potency factors, ingestion reference doses	Surrogate chemical or most toxic chemical in class
Ecological benchmark	Ecological risk	Water quality criteria, soil quality criteria, lowest affect dose for population endpoint	Surrogate chemical or estimation programs like ECOSAR for aquatic life
Molecular weight (and chemical structure)	Source, Surface water	-	None
Partition coefficients	Multiple	OPPTS 835.1220 (OECD 106) OPPTS 830.7550 (OECD 107)	EPI Suite, SPARC, and established estimation equations
Water Solubility	Source, Water modules	OPPTS 830.7840 (OECD 105)	EPI Suite, SPARC
Critical pressure	Source	-	EPI Suite, SPARC
Critical temperature	Source	-	EPI Suite
Boiling point	Source	OPPTS 830.7220 (OECD 103)	EPI Suite, SPARC
Vapor pressure coefficients	Source	OPPTS 830.7950 (OECD 104)	EPI Suite, SPARC
Henry's Law Constant	Multiple	-	EPI Suite and established estimation equations
Diffusivity in air	Source	-	SPARC
Diffusion coefficient in water	Source, Groundwater	-	SPARC
Ionization equilibrium constant (requires acid base designation)	Multiple	OPPTS 830.7370 (OECD 112)	SPARC
Soil degradation rate*	Watershed, Source	OPPTS 835.3110 (OECD 301)	EPI Suite
Surface water degradation rate*	Surface water	OPPTS 835.4100 (OECD 307)	EPI Suite
Groundwater degradation rate*	Groundwater	OPPTS 835.6100	EPI Suite
Bioconcentration factors	Aquatic food web	OPPTS 850.1850 OPPTS 850.1730 (OECD 305)	EPI Suite
Bioaccumulation factors	Terrestrial food web	OPPTS 850.4800 OPPTS 850.6200 (OECD 207)	Methods available for plants (BTFs) and worms, but not well developed for other prey
Biotransfer factors	Farm food chain	OPPTS 870.8320 OPPTS 870.8340	Available methods for plant uptake, beef, and dairy

* An overall media-specific degradation rate is typically a function of degradation rate associated with specific biotic and abiotic processes such as biodegradation (aerobic and anaerobic), hydrolysis, photolysis, etc. Note that, for screening purposes, degradation rates are sometimes assumed to be zero or very low (using a highly persistent organic chemical as a surrogate) to support a conservative model simulation.

The risk assessment procedure follows a tiered approach where conservative assumptions and model predictions are initially used and input data are refined at higher tiers. Exposure models are generally used to determine the PEC at lower tiers, whereas field monitoring data is used at higher tiers. The PNEC is determined by dividing the effects concentration for the most sensitive relevant species for the compartment of interest by an assessment factor that reflects the uncertainty of extrapolating from laboratory to field conditions. The initial tier of the assessment may use quantitative structure activity relationship (QSAR) predictions for estimating toxicological effects. The next tier would involve conducting laboratory acute or chronic toxicity tests. For terrestrial assessments, laboratory tests at lower tiers may use artificial or natural substrate. Higher tiered testing may include aged-residue studies, model ecosystem tests, or field studies.

3.3.1 Exposure Assessment

The amount and frequency of biosolids amendment to soil depends on whether pasture or crop land is assessed. For the local PEC calculation, biosolids application is assumed to occur annually for 10 consecutive years (CEC, 1991a; CEC, 1991b; CEC, 2003). For biosolids applied to crop land, the incorporation depth is assumed to be 20 cm and the rate of application is 0.5 $kg/m^2/y$ (5 $Mg\ ha^{-1}\ y^{-1}$). For biosolids applied to pasture land, the incorporation depth is assumed to be 10 cm and the rate of application is 0.1 kg/m^2/yr (1 Mg ha^{-1}y^{-1}). Chemical exposure is assessed 30 days after application, which is the time required before crops can be planted in the amended field or animals are allowed on amended pasture land. The assessment further assumes that the chemical concentration reaches steady-state over a period of 10 years and that the chemical is completely bioavailable. This approach ignores the processes of aging or humification (Hatzinger and Alexander, 1995; Luthy et al., 1997), which could be important when assessing biodegradation and toxicological effects.

Chemical loss from the soil is assessed using a simple completely-mixed box model that includes biodegradation, volatilization, and leaching as loss processes. These processes begin immediately after the biosolids are amended to soil, though exposure concentrations are assessed after 30 days. The estimation of chemical loss by volatilization depends on several physical-chemical parameters including vapor pressure, water solubility, and soil-water partitioning. Volatilization of biosolids-borne TOrCs is expected to be minor in amended soil. Chemical leaching is governed by chemical soil-water partitioning and the amount of rain water. Biodegradation rates are typically extrapolated from aquatic biodegradation tests (e.g., OECD 301 Ready Biodegradability), though the extrapolation from aquatic media to soil systems is poorly known. Moreover, the aquatic ready biodegradation test does not provide kinetic information and loss rates are assigned based on how well the compound performs in the test. To reduce uncertainties in extrapolation, it is always best to measure biodegradation in soil systems. A list of acceptable test methods for assessing the fate and exposure of TOrCs is presented in Table 3-9.

Table 3-9. Acceptable Test Methods for Assessing Terrestrial Fate and Exposure.

Test Description	Test Method
Physical-Chemical Properties	
Vapor pressure	OECD 104
Water solubility	OECD 105
Adsorption-desorption using batch equilibrium method	OECD 106
Partition coefficient (n-octanol/water): shake flask method	OECD 107
Hydrolysis as a function of pH	OECD 111
Dissociation constant in water	OECD 112
Partition coefficient (n-octanol/water), HPLC method	OECD 117
Estimation of adsorption coefficient (K_{oc}) in soil, HPLC method	OECD 121
Degradation	
Ready biodegradability, CO_2 evolution	OECD 301B
Inherent biodegradability, modified MITI	OECD 302C
Simulation test – aerobic sewage treatment	OECD 303A
Inherent biodegradation in soil	OECD 304A
Aerobic and anaerobic transformation in soil	OECD 307
Leaching in soil columns	OECD 312

3.3.2 Effects Assessment

Toxicity data are needed for earthworms, plants, and microorganisms, though data from other taxa are acceptable. For most chemicals, toxicological data for soil organisms are limited. In the absence of measured data, the EU TGD suggests that an equilibrium partitioning method be used to extrapolate soil toxicity from aquatic toxicity (CEC, 2003). Higher tiered tests that use natural substrates or controlled field studies may be used if lower tiered results indicate possible risk. A description of the tiered testing approach for conducting an effects assessment is provided in the EU TGD. A select list of acceptable test methods for assessing terrestrial toxicological effects is presented in Table 3-10.

Table 3-10. Acceptable Test Methods for Assessing Terrestrial Effects.

Test Description	Test Method
Terrestrial Vertebrates	
Avian acute oral toxicity	OECD 223
Avian short-term dietary toxicity	OECD 205
Avian reproduction	OECD 206
Terrestrial Invertebrates	
Honeybees acute oral toxicity	OECD 213
Honeybees acute contact toxicity	OECD 214
Soil Macroorganisms	
Earthworm acute toxicity	OECD 207
Earthworm reproduction	OECD 222
Earthworm field	ISO 11268
Soil Microorganisms	
Nitrogen transformation	OECD 216
Carbon transformation	OECD 217
Terrestrial Plants	
Seedling emergence and seedling growth	OECD 208
Vegetative vigor	OECD 227

Assessment factors are used in the effects assessment to address experimental uncertainty, species sensitivity, acute to chronic ratio extrapolations, life stage sensitivity, mode of action, and laboratory to field extrapolation. The soil PNEC is then determined by dividing the lowest measured effects value by an appropriate assessment factor. The assessment factors in the EU TGD are presented in Table 3-11 (CEC, 1996).

Table 3-11. Assessment Factors for Most Sensitive Species to Determine PNEC in Soil.

Toxicity Test	Assessment Factor
Acute toxicity (plants, earthworm, microorganisms)	1000
1 chronic NOEC (one trophic level)	100
2 chronic NOECs (two trophic levels)	50
3 chronic NOECs (three trophic levels)	10
Field or model ecosystem	1 - 5 (case by case)

3.3.3 Risk Characterization

The European Industry Council developed a simulation model - European Union System for the Evaluation of Substances (EUSES) - to facilitate the risk assessment process of chemicals in accordance with the EU TGD (Lijzen and Rikken, 2004). EUSES is an integrated-modeling decision-support tool that uses the multimedia fate model SimpleBox (den Hollander et al., 2004) to determine the distribution and fate of chemicals in the environment. It is capable of simulating three scales for Europe: local, regional, and continental scale. Based on the quantity of emissions in air, water, or soil environments, the model calculates steady-state concentrations in each compartment based on physical-chemical parameters. The model assumes equilibrium partitioning among the three compartments and considers degradation processes within each environmental compartment. Removal of contaminants during wastewater treatment and biosolids concentrations are predicted using the SimpleTreat model (Struijs, 1996). The output from EUSES is a risk characterization ratio (RCR) for air, surface water, sediment, soil, and biota. The RCRs are calculated by dividing the PEC by the PNEC. A substance is of potential concern if the RCR >1.

EUSES is a complex, integrated multimedia model that requires a significant amount of chemical-specific data. Knowing the amount of chemical entering the sewerage system and its removal during wastewater treatment is important for determining biosolids concentrations. Once a chemical is released into the soil environment, parameters such as soil biodegradation rates, soil-water partition coefficients, and bioaccumulation factors are needed for assessing the chemical's fate and effects. While volatilization could be a key parameter for highly volatile compounds, this parameter is typically more important for predicting removals during wastewater treatment than for assessing fate in biosolids-amended soil.

Some of the parameters in EUSES have a linear effect on the RCR, whereas other parameters have a non-linear effect. For example, the amount of chemical entering a wastewater treatment plant has a linear response, thus a doubling of the down-the-drain load will double the RCR. The PNEC also has a linear effect on the RCR, thus a doubling the PNEC will also double the RCR. On the other hand, fate processes of sorption, volatilization, and biodegradation have non-linear effects on the RCR. Therefore, care should be taken when predicting exposure concentrations with these fate parameters because their non-linear response on the RCR causes higher uncertainty in the assessment. This caution is particularly true when QSARs are used to predict sorption from log K_{ow} and volatilization (K_{aw}) from water solubility and vapor pressure values, which may also be predicted. Biodegradation rates in EUSES are set depending on if the chemical is classified as non-biodegradable, inherently biodegradable, or readily biodegradable using standard OECD test methods.

3.3.4 Data Needs

A Tier 1 assessment can be conducted with EUSES using only "base set" information required by the EU. For exposure assessments, the minimum data set for EUSES is octanol-water partition coefficient, water solubility, vapor pressure, and ready biodegradation. For the effects assessment, only aquatic acute toxicity values are needed for three trophic levels (e.g., algae, daphnia and fish) to predict the no effect concentrations. Extrapolation techniques (e.g., QSARs) can be used to assess risk at the Tier 1 level, though larger assessment factors are applied due to higher uncertainty in these predictions. If the RCR > 1 after a Tier 1 assessment, then higher tiered laboratory tests in soil systems can be used to refine the terrestrial assessment. A select list of acceptable fate and toxicity tests for soil systems is presented in Table 3-9 and Table 3-10, respectively.

3.4 International Life Sciences Institute – Europe

The Environmental and Health Task Force of the European branch of the International Life Sciences Institute (ILSI Europe) developed a conceptual framework for deriving quality standards of TOrCs in biosolids-amended soils (Schowanek et al., 2004). The methodology is based on a tiered assessment approach that compares predicted exposure concentrations to the predicted no effect concentrations (i.e., PEC to PNEC). For human risk, a comparison of acceptable daily intake (ADI: mg/kg body weight day) to exposure (mg/kg body weight day) is made. The approach is consistent with the EU TGD for environmental risk assessment of chemicals in the soil compartment and the US EPA Part 503 regulations.

The conceptual framework describes a stepwise procedure to derive biosolids quality standards based on risk assessment procedures. This framework is similar to the "back-calculating" mode used in the U.S. EPA approach.

Step 1 – Effects Assessment

An effects assessment is conducted to determine a PNEC based on relevant endpoints.

Step 2 – Exposure Assessment

The soil PNEC defines the maximum acceptable exposure level or soil PEC. The Sludge Quality Standard (SQS) is then derived from the soil PEC using typical agronomic application rates. This procedure provides for a maximum biosolids PEC, which can be expressed as a mass flux to the soil compartment rather than a biosolids concentration.

Step 3 – Validation

Repeated applications of biosolids to soil require defining the time horizon for which the assessment is valid. The approach ensures that there is no build up of a chemical in soil above the PNEC value. The initial soil contaminant concentration is assumed to be zero (prior to biosolids amendment) and accumulation of the chemical is due only to addition of biosolids to the soil. Soil mass balance models are used to assess the risk of accumulation over time (Andersen, 2001; Jones and Stevens, 2002).

3.4.1 Exposure Assessment

The concentration of a TOrC in biosolids depends on several factors such as: 1) sorption, which is often controlled by hydrophobicity; 2) volatilization, which is highly dependent on Henry's Law constant; and 3) degradation processes, which can either be abiotic (hydrolysis or photolysis) or biotic (biodegradation or biotransformation). The SimpleTreat model is used to predict TOrC concentrations in biosolids. Following land application, TOrCs are subject to chemical or biological processes and physical transfer processes (leaching, runoff, and volatilization).

The potential for TOrCs to leach to groundwater following biosolids application to agriculture land is assessed using mathematical models, data generated from laboratory leaching columns (e.g., OECD 312: Leaching in Soil Columns), or lysimeters in the field. The mobility of a TOrC in soil depends on its physical-chemical properties (water solubility, vapor pressure, octanol-water partitioning, organic carbon-water partitioning, Henry's Law constant) and persistence (i.e., half-life). Data for all of these parameters can be used as input to the soil fate and transport models. Caution should be taken, though, when using the organic-carbon partitioning approach for ionizable compounds or compounds with a low log K_{ow}.

3.4.2 Effects Assessment

The effects assessment considers a variety of species including soil fauna and flora that are directly exposed to TOrCs, domestic and wildlife animals that consume plants or soil biota from the amended soil, and aquatic organisms in nearby surface waters that may be exposed from leaching and runoff. Human and domestic animals may ingest TOrCs from onsite groundwater wells or nearby surface waters. Soil ingestion from unwashed vegetables or from pica behavior is also considered for human health effects.

The soil PNEC for a TOrC is derived from ecotoxicity tests. Soil organisms that represent the ecological function of the ecosystem (e.g., microbial respiration) are also considered. Once an effect concentration is determined for individual species, the PNEC is derived by using assessment factors described in the EU TGD (CEC, 2003). For data rich chemicals, a statistical extrapolation technique can be used that derives a PNEC from the distribution of chronic no-effect single species test data (Suter, 1993; CEC, 2000; Posthuma et al., 2002). The technique seeks to protect 95% of the species in the compartment of interest.

The effects of a chemical to microbial populations are evaluated by using a battery of tests. These tests include microbial numbers (Garland and Mills, 1991; Lawlor et al., 2000), microbial biomass (Khan and Scullion, 1999; Albiach et al., 2000), microbial metabolic activity (Tate and Jenkinson, 1982; Baird and White, 1985; White, 1995), respirometry (Khan and Scullion, 1999; Saviozzi et al., 1999), and molecular biology techniques (Moran et al., 1993). A microbial PNEC is derived from the results of these tests.

Numerous test methods are available to assess the toxicity of TOrCs to soil invertebrates and plants. A comprehensive list of test species and protocols for both soil invertebrates and plants is provided in the EU TGD (CEC, 2003). A select list of acceptable test methods for assessing terrestrial toxicological effects is presented in Table 3-10.

The potential exposure to birds and mammals is assessed using BAFs derived from earthworm studies (Connell and Markwell, 1990). If the predicted concentration in the food exceeds the PNEC, then secondary poisoning via ingestion of prey is considered a critical pathway. The EU TGD provides more detail on how biomagnification and secondary poisoning are assessed.

3.4.3 Risk Assessment

The risk assessment of TOrCs following biosolids amendment to soil is based on the PEC to PNEC risk ratio. This approach requires careful consideration of soil types used in the evaluation (organic carbon content, clay content, pH, etc.) because soil properties can affect bioavailability of the test compound and processes such as aging and humification. The approach also uses realistic test systems by dosing the TOrCs to soil in a biosolids matrix rather than an aqueous solution. The soil PEC estimation follows methods outlined in the EU TGD (CEC, 2003).

Indirect exposure via bioaccumulation and biomagnification is also considered. The approach is described in the EU TGD and EUSES software documentation (Van de Meent et al., 1995; CEC, 2003; Lijzen and Rikken, 2004). Dietary intake from drinking water, fish, food crops, milk, and meat are combined to calculate a total daily uptake of the chemical. The approach relies on linear transfer coefficients or biotransfer functions based on lipophilicity (log K_{ow}). Estimated exposure concentrations are compared to ADI values or reference dose (RfD) values. The assessment is based on daily exposure over a 70 year lifetime that will not cause adverse effects.

The level of exposure by direct soil ingestion is determined from the concentration of chemical in soil, the amount of soil consumed, the body weight of the consumer, the percentage of chemical absorbed by the body, and the duration of exposure. The estimated exposure is then compared to the ADI or RfD values to assess the risk. For certain chemicals, the chronic lifetime exposure assessment may not be protective of children during acute soil pica events. Children that exhibit soil pica behavior can ingest large amounts of soil, which may result in the risk of acute toxicity. The acute assessment assumes that a 20 kg child ingests 60 g of biosolids-amended soil (single dose scenario).

Assessing respiratory exposure to humans for volatile TOrCs can be complex and often relies on expert judgment. Multimedia models may also help refine the assessment. These models rely on physical-chemical parameters such as Henry's Law constant, surface sorption, and diffusivity. Existing occupational exposure models can be used to perform the risk assessment (U.S. EPA, 1992); however, in most cases this is not a critical exposure pathway for TOrCs in biosolids amended soil.

A case study example of this methodology has been published for linear alkylbenzene sulfonate (Schowanek et al., 2007). This chemical was chosen for the case study because it has a considerable amount of test data available. The study provides an excellent overview on how the ILSI approach can be implemented.

3.4.4 Data Needs

The data needs for the ILSI approach are very similar to those summarized in the EU TGD approach (section 3.2.4) and are not repeated here.

3.5 Data Sources and Estimation Tools

The assessment techniques described above require significant data inputs. There are several sources of data to help conduct a risk assessment, but when no data are available QSAR models are used to estimate the needed data. Data sources and estimation tools for conducting risk assessments are provided below.

3.5.1 Data Sources

Information on data sources for conducting risk assessments is presented in Table 3-12.

Table 3-12. Data Sources for Conducting a Terrestrial Risk Assessment.

Source	Website
The Human and Environmental Risk Assessment (HERA) project provides comprehensive risk assessments for several chemicals in European household cleaning products. The HERA project is a voluntary collaboration among formulations and suppliers of household cleaning products.	http://www.heraproject.com/RiskAssessment.cfm
The Organization of Economic Co-operation and Development (OECD) has complied information on existing high production volume chemicals.	http://www.oecd.org/document/55/
EPA Office of Pollution Prevention and Toxics (OPPT) has also compiled high production volume reports similar to OECD.	http://www.epa.gov/hpv/
The U.S. EPA maintains a chemical toxicity database relevant to conducting ecological risk assessments. Information is tabulated for aquatic species, terrestrial animals (primarily wildlife species), and terrestrial plants and includes lethal and sublethal toxic effects data.	http://www.epa.gov/ecotox/
Environment Canada's existing substance division has complied data on chemicals that are regulated under the Canadian Environmental Protection Act of 1999.	http://www.ec.gc.ca/substances/ese/eng/esehome.cfm

3.5.2 Estimation Tools

The U.S. EPA OPPT developed PBT Profiler for estimating persistence, bioaccumulation, and toxicity of chemicals in the absence of experimental data (http://www.pbtprofiler.net/). The EPI Suite tool was also developed for estimating physical-chemical parameters and environmental fate of chemicals (http://www.epa.gov/oppt/exposure/pubs/episuite.htm). A list of some of the chemical properties and fate parameters in EPI Suite is presented in Table 3-13.

The EU developed a tool (OECD QSAR Application Toolbox) that can be used to search for available experimental data, identify analogues for a chemical, group chemicals by mode of action or structural similarity, fill data gaps using read-across, perform trend analysis or QSAR model analysis, and fill data gaps for a chemical using a QSAR model library. The toolbox can be downloaded and is available free of charge (http://www.oecd.org/document/23/0,3343,en_2649_34379_33957015_1_1_1_1,00.html).

Table 3-13. Chemical Property and Fate Programs in EPI Suite.

Program	Description
AOPWIN	Atmospheric oxidation
BCFWIN	Bioconcentration factor (BCF)
BIOWIN	Biodegradability
HENRYWIN	Henry's Law constant
HYDROWIN	Aqueous hydrolysis
KOWWIN	Octanol-water partitioning coefficient
MPBPVP	Melting point, boiling point, vapor pressure
PCKOC	Soil sorption coefficient (K_{oc})
WSKOW	Water solubility from log K_{ow}
WATERNT	Water solubility from fragments
STPWIN	Removal in activated sludge treatment
Level III	Transport/distribution by fugacity
WVOLWIN	Volatilization from water

3.6 Summary of Data Needs for Terrestrial Risk Assessments

The accuracy of a risk assessment depends greatly on the quality of the input parameters. Because only a few TOrCs have robust data sets for assessing risk, a major issue for risk assessors is deciding on how best to deal with the many chemicals with limited data. In particular, data are needed for the ecological pathways of most chemicals. Some of the necessary information can be predicted, but uncertainties in the predictions may ultimately require laboratory and field testing. The key information needed for conducting and refining a risk assessment is presented below.

3.6.1 Emissions to Soil

Accurate estimates of chemical additions to soil environments are critical for assessing risk. The U.S. EPA approach is to analyze a large number of biosolids for contaminants of interest. The 95th percentile concentration is then obtained from the distribution. In contrast, the European approach is to predict the biosolids concentration using the mathematical model SimpleTreat. Estimates of down-the-drain tonnages of the chemical, and information on the sorption, biodegradation and volatilization of the chemical, are needed for accurate predictions of biosolids contaminant concentrations. The amount of biosolids amended to soil, frequency and length of time between amendments, and other management practices specified by existing regulations (set-back distances, slope limitations, best management practices, etc.) are all critical in assessing ultimate exposure concentrations in the soil environment.

3.6.2 Partitioning in Soil

Accurate estimates of chemical partitioning in soil (sorption/desorption) are critical for assessing risk. This process is important when assessing a chemical's bioavailability, toxicity, bioaccumulation, biodegradation, volatilization, and mobility to groundwater. Ideally the sorption of a chemical is measured in several soil types. Understanding the kinetics of adsorption and desorption, aging and facilitated transport may also be important for assessing a chemical's ultimate fate in soil. Detailed laboratory studies would be needed to assess these fate and transport mechanisms.

3.6.3 Degradation in Soil

Accurate estimates of the degradation of a chemical in biosolids-amended soil are critical for conducting a meaningful risk assessment. Biodegradation rates are difficult to predict and are best measured at relevant concentrations in realistic laboratory or field soil systems. The EU TGD recommends the extrapolation of soil biodegradation rates from aquatic ready biodegradation test data when measured soil data are not available. This approach is problematic and requires caution if used. Also, for some chemicals abiotic degradation may be an important loss mechanism. An example of this mechanism is the abiotic clay-catalyzed hydrolysis of polydimethylsiloxanes in soil (Xu et al., 1998; Traina et al., 2002).

3.6.4 Ecological Effects in Soil

Accurate predictions of the ecological effects of a chemical in soil environments are fundamental to assessing terrestrial risk. Standard test methods exist for assessing the effects to terrestrial organisms (OECD and OPPTS), but factors such as spiking methods (via biosolids), soil composition (% sand, silt, clay and organic matter), study length, and realistic TOrC concentrations should be considered carefully. Greater standardization will ensure consistency among risk assessments. Without proper validation, extrapolation techniques from different systems (e.g., aquatic to soil) should be used with caution. For example, the equilibrium partitioning approach presented in the EU TGD makes several assumptions that do not follow the original theory by Di Toro et al. (1991). The Di Toro et al. approach predicts effects to sediment-dwelling organisms by comparing sediment pore water concentrations to water column species. While it may be appropriate to predict effects to some sediment-dwelling organisms from water column effects data, it is less clear if these same extrapolation techniques can be extended to soil-dwelling organisms. Additional data is also needed for assessing the ecological effects of mixtures in the soil environment. Current approaches only assess risk on a chemical by chemical basis. This approach neglects the possibility of synergistic or antagonistic effects of multiple chemicals in the soil environment. Experimental data would be needed to address mixtures in terrestrial ecological systems.

3.6.5 Bioaccumulation in Soil

Accurate estimates of bioaccumulation and biotransfer factors are critical for conducing ecological and human health risk assessments. In the absence of measured data, model predictions are often used or a BAF of 1 is assumed. The methods for predicting biological-uptake of organic compounds require a log K_{ow} value and utilize relationships that may not be applicable for all TOrCs. Moreover, these models do not take into account transformations that may take place within the organism for some chemicals. Thus, measured data is need for accurate BAF and BTF values. This type of data is also important for assessing food web responses.

3.7 Summary of Modeling Needs for Risk Assessments

3.7.1 Sorption in Soil

Current risk models assume equilibrium partitioning (i.e., K_{oc}) for predicting sorption of TOrCs in soil. This approach may be appropriate for hydrophobic compounds, but not for compounds that ionize under ambient pH conditions. Therefore, a different modeling approach is needed to better describe the sorption of ionogenic chemicals in soil systems.

Current risk models assume instantaneous equilibrium between the soil and aqueous phases. However, rates of adsorption and desorption could play an important role in bioavailability, biotransformation, and transport for some organic compounds (particularly when TOrCs are introduced to soil in a biosolids matrix). Therefore, model formulations are needed to better describe the sorption of organic compounds that are kinetically controlled.

Another potential need for current risk models is the sorption of TOrCs to colloidal material, which is needed to model facilitated transport in vadose and groundwater systems.

3.7.2 Biotransformation in Soil

Current risk models assume first-order loss of a chemical in the soil compartment. For some organic compounds, there may be other reaction orders that more appropriately describe their loss in soil (e.g., zero-order, second-order, or three-half-order). For example, a three-half-order model may better describe the slow mineralization phase often observed in the later portion of mineralization curves. The reaction order may also be affected by factors such as: 1) the compound is cometabolized, not metabolized; 2) the rate of desorption from soil is slower than the biotransformation rate; or 3) humification is taking place over time. Therefore, there is a need to better understand the bias and uncertainty associated with assuming first-order loss for those TOrCs that are better described by other kinetic models, and if warranted incorporate different kinetic degradation models in the risk assessment framework. This capability would allow risk assessors more options to accurately represent the process of chemical loss in risk models.

3.7.3 Transformation in Plants

Current risk models do not adequately account for translocation and biotransformation of organic compounds in plants. Modeling the translocation of TOrCs into edible portions of plants is of particular importance, though biotransformation by plant tissue may be an important loss mechanism for some TOrCs. Therefore, there is a need to evaluate the importance translocation and transformation of TOrCs in plants, and if warranted incorporate these processes into the plant bioaccumulation model.

3.7.4 Biosolids Concentrations

Accurate predicted concentrations of TOrCs in biosolids are needed when no measured data are available. The U.S. EPA risk model currently does not have the capability to predict TOrC concentrations in biosolids because it either uses measured concentrations of TOrCs in biosolids as an input to the soil compartment or it uses a "limits based" approach (backward-calculating mode) for estimating acceptable concentrations of TOrCs in biosolids. On the other hand, the European EUSES risk model uses SimpleTreat to predict biosolids TOrC concentrations (forward-calculating mode). Existing wastewater treatment plant models (e.g., SimpleTreat, ASTreat, ToxChem) have demonstrated a good ability to predict effluent concentrations of TOrCs, but these models have not been adequately validated for predicting TOrC concentrations in biosolids. A combination of laboratory and field studies would be needed to evaluate the accuracy of these models. Therefore, there is a need to evaluate existing wastewater treatment plant models for their ability to accurately predict concentrations of TOrCs in biosolids when the risk assessment models are used in a forward-calculating mode.

3.7.5 Sensitivity Analysis

A rigorous sensitivity analysis is needed to quantify the importance of input parameters on the risk model output. This type of an analysis could help focus the data gather effort and potentially reduce the number of parameters needed in performing a biosolids risk assessment.

3.7.6 Field Validation

The current U.S. EPA approach has not been thoroughly field evaluated. Several well designed field studies are needed to provide confidence in the risk assessment approach. The field evaluations should be holistic in nature (including both terrestrial and aquatic pathways) and designed such that the most important driving variables (e.g., soil pH, rainfall) in the risk assessment are represented. Such an evaluation would provide assurance that the modeling approach is appropriate.

3.7.7 Tiered Assessment Approach

Currently the U.S. EPA uses a probabilistic risk assessment approach, which requires a significant amount of data to perform. Because of the data requirements for this higher tier approach, it cannot be easily used for those organic compounds with limited data. Therefore, a lower (screening) tier assessment approach is needed to assess TOrCs with little data. A screening tier approach could help the U.S. EPA determine which exposure pathways are most important for a given TOrC and what are the critical data gaps for those pathways.

CHAPTER 4.0

OCCURRENCE OF TRACE ORGANIC CHEMICALS IN MUNICIPAL BIOSOLIDS

4.1 Introduction

The reported widespread occurrence of numerous trace organic chemicals (TOrCs) in municipal biosolids was one of the main motivations for this study. As discussed in Chapter 2.0, many TOrCs have been detected and quantified in biosolids, but whether chemical presence constitutes significant risk is a question that remains unanswered for many TOrCs. This chapter summarizes the findings of two major surveys of TOrCs in biosolids conducted by US government laboratories as well as the findings from smaller surveys of biosolids and municipal sludge conducted by various investigators. Only occurrence data for the TOrCs included in this study (Table 2.1) are summarized. Limited occurrence data for TOrCs excluded from this study (due to low occurrence or infrequent detection) are provided in Table 2.3 (Chapter 2.0).

Available occurrence data were used to prioritize TOrCs for inclusion in this study, but data gaps do exist with respect to TOrC occurrence. For example, not all the TOrCs of concern have been analyzed for in large surveys, suggesting that the average concentrations reported here may be biased by small sample sizes. Second, this study was not intended to evaluate what TOrCs *might potentially occur* in biosolids and therefore *might potentially pose a risk* to humans and the environment. Such an effort has recently been conducted with respect to household chemicals in wastewater (Drewes et al., 2008), but a similar effort for biosolids was beyond the scope of the present study. In addition, some of the studies included in this chapter focused on occurrence in sewage sludge as opposed to finished biosolids. Declining concentrations of some of the targeted TOrCs during storage of biosolids have been documented (Wu et al., 2008), suggesting that occurrence data relying solely on sewage sludge may overestimate the levels of some TOrCs present in the biosolids actually applied to soils. The collective available occurrence data for the high priority TOrCs in sewage sludge (and biosolids) are fairly substantial. This chapter provides a summary of the data availability and identifies data gaps that may still need to be filled before meaningful risk assessments could be conducted.

As discussed in Chapter 3.0, the maximum concentrations of TOrCs in biosolids-amended soils can be estimated from assumptions made about the mixing depth (typically 15 to 20 cm) and data (or estimates) of the biosolids application rate and the concentrations of TOrCs in the biosolids. These estimates of soil concentrations are maximal values, as dissipation processes may lead to substantial reductions in concentrations both before and after the biosolids are incorporated into the soil. For example, Wu et al. (2008) found the levels of some pharmaceuticals and personal care products (PPCPs) to dissipate under condition meant to mimic storage prior to land application, suggesting that levels of TOrCs measured in sewage sludge may overestimate the concentrations in biosolids at the time of application. In addition, the length of time between biosolids application and soil sampling further complicates interpretation of field data. Given these variables, a tabulation of concentrations of TOrCs in biosolids-amended soils would be of limited use. Nevertheless, several studies have detected many of the targeted TOrCs in biosolids-amended soils, and a brief overview of these studies is provided.

Of utmost importance in occurrence surveys for TOrCs in sewage sludge or biosolids is the use of appropriate analytical methods. Matrix effects and inadequate extraction and/or clean-up procedures can lead to significant errors in reported concentrations. Fortunately, the two national surveys that are the focus of this chapter employed stable isotope dilution mass spectrometry, which generally provides the most accurate and reliable quantitative data. However, the analytical methods employed by various investigators to document the presence of the target TOrCs in biosolids and biosolids-amended soils were not subject to detailed scrutiny (nor were the analytical methods employed by investigators discussed in subsequent chapters). In general, methods relying on mass spectrometry and employing stable isotope dilution to correct for recoveries and matrix effects are the most desirable. Analytical methods are constantly evolving, resulting in greater sensitivity. Thus, a non-detect (ND) in an earlier survey may become a measurable concentration in a later survey if the sensitivity of the method improved. The potential variability in the methods employed and their sensitivities must be kept in mind, particularly when considering occurrence data developed by individual investigator studies.

4.2 National Surveys of Trace Organic Chemicals in Municipal Sludge

The two most comprehensive and recent surveys of TOrCs in municipal U.S. biosolids were conducted by the U.S. EPA (U.S. EPA, 2009i) and the U.S Geological Survey (USGS) (Kinney et al., 2006). Occurrence data from these surveys for the TOrCs included in this study are provided in Table 4-1. Additional occurrence data for TOrCs included in this study, but not included in either the USGS survey or the Targeted National Sewage Sludge Survey (TNSSS), are also provided in Table 4-1. The two surveys were the most comprehensive in terms of sample size and analytical coverage. Furthermore, for TOrCs included in both surveys, the range of concentrations (dry weight basis) reported in Table 4-1 reflects the full range of both surveys. In other words, the minimum concentration reported is the minimum concentration from both surveys, and the maximum concentration reported is the maximum from both surveys. When available from both surveys, the *maximum* mean and median concentrations were included to provide a conservative data set. For TOrCs not included in either the USGS survey or the TNSSS, the same approach for data from other studies was used to define the full range of concentrations and conservative estimates of mean and median concentrations.

4.2.1 Targeted National Sewage Sludge Survey

The most comprehensive occurrence survey for many of the TOrCs included in this study was the TNSSS conducted in late 2006 and early 2007 (U.S. EPA, 2009i). Sewage sludges were collected from treatment plants representing a variety of biosolids management practices, including land application. The survey included 74 large facilities accounting for approximately 94% of the wastewater flow in the U.S. The TNSSS examined the occurrence of 145 analytes, 40 of which are TOrCs included in the present study (20 high priority TOrCs).

Table 4-1. Occurrence of Trace Organic Chemicals Included in This Study.

Chemical	Chemical Class	Use	Max Mean (µg/kg)	Max Median (µg/kg)	Range (µg/kg)	Data Source
		High Priority				
4-Epitetracycline	PPCPs	Antibiotic	1,135	620	41 – 4,380	1
Cimetidine	PPCPs	Antacid	1,332	171	4 – 8,330	1,2
Ciprofloxacin (CIP)	PPCPs	Antibiotic	10,501	5,367	75 – 40,800	1,2
Doxycycline (DTC)	PPCPs	Antibiotic	877	5,367	34 – 5,090	1
Galaxolide (HHCB)	PPCPs	Fragrance Material	37,600	1,100	13 – 86,000	2, 3
Mestranol (MeEE2)	Steroidal Chemicals	Synthetic Hormone	NA	NA	NA	
Miconazole	PPCPs	Antifungal	1,239	207	ND – 9,210	1,2
Ofloxacin	PPCPs	Antibiotic	8,573	3,113	ND – 58,100	1,2
Tetracycline (TC)	PPCPs	Antibiotic	1,278	630	ND – 5,270	1
Triclocarban (TCC)	PPCPs	Antimicrobial	39,433	21,677	187 – 441,000	1
Triclosan (TCS)	PPCPs	Antimicrobial	16,097	3,862	ND – 133,000	1, 2
17α-Ethinyl estradiol (EE2)	Steroidal Chemicals	Synthetic Hormone	25	25	9 – 30	1
BDE 28	BFRs	Fire Retardant	15	9	2 – 160	1
BDE 47	BFRs	Fire Retardant	709	570	60 – 5,000	1, 2
BDE 85	BFRs	Fire Retardant	278	23	3 – 150	1
BDE 99	BFRs	Fire Retardant	716	575	64 – 4,000	1
BDE 100	BFRs	Fire Retardant	150	120	13 – 1,100	1
BDE 138	BFRs	Fire Retardant	11	7	ND – 40	1
BDE 153	BFRs	Fire Retardant	68	54	9 – 410	1
BDE 154	BFRs	Fire Retardant	60	47	8 – 440	1
BDE 183	BFRs	Fire Retardant	17	10	2 – 120	1
BDE 209	BFRs	Fire Retardant	2,181	1,163	ND – 17,000	1
Dimethyl TBBPA	BFRs	Fire Retardant Metabolite	ND	ND	ND	4
Hexabromocyclododecane (HBCD isomers)	BFRs	Fire Retardant	1,633	1,401	ND – 9,120	5
TBBPA	BFRs	Fire Retardant	78	79	0.031 – 600	4, 5
6:2/8:2diPAPs	PFCs and Precursors	Surface Coatings	55	28	9 – 160	6
6:2diPAPs	PFCs and Precursors	Surface Coatings	164	91	55 – 590	6
8:2/10:2diPAPs	PFCs and Precursors	Surface Coatings	198	160	4 – 550	6
8:2diPAPs	PFCs and Precursors	Surface Coatings	186	98	12 – 860	6
10:2/12:2diPAPs	PFCs and Precursors	Surface Coatings	112	38	4 – 600	6
10:2diPAPs	PFCs and Precursors	Surface Coatings	78	63	28 – 220	6
PFHxA	PFCs and Precursors	Surface Coatings	9	7	1 – 18	7
PFHpA	PFCs and Precursors	Surface Coatings	7	7	ND – 10	7
PFOA	PFCs and Precursors	Surface Coatings	2,023	1,721	ND – 4,780	8
PFNA	PFCs and Precursors	Surface Coatings	5	NA	ND – 10	9
PFDA	PFCs and Precursors	Surface Coatings	52	46	1 – 91	10, 11
PFUnDA	PFCs and Precursors	Surface Coatings	60	40	<25 – 115	11
PFDoDA	PFCs and Precursors	Surface Coatings	7	4	<2 – 20	12
PFTriDA	PFCs and Precursors	Surface Coatings	8	8	<2 – 10	7
PFTeDA	PFCs and Precursors	Surface Coatings	6	5	<2 – 20	7
FOSA	PFCs and Precursors	Surface Coatings	14	NA	<3 – 21	12
FOSAA	PFCs and Precursors	Surface Coatings	14	7	ND – 62	9
N-EtFOSAA	PFCs and Precursors	Surface Coatings	185	130	61 – 544	9

NA = Not available; ND = not detected

Data Sources:

1 (U.S. EPA, 2009i)
2 (Kinney et al., 2006)
3 (Difrancesco et al., 2004)
4 (Sellstrom and Jansson, 1995)
5 (Morris et al., 2004)
6 (D'Eon et al., 2009)
7 (Yoo et al., 2009)
8 (Guo et al., 2008)
9 (Higgins et al., 2005)
10 (Bossi et al., 2008)
11 (Sinclair and Kannan, 2006)
12 (Loganathan et al., 2007)
13 (Poiger et al., 1998)
14 (Kupper et al., 2004)
15 (Herren and Berset, 2000)
16 (Tan et al., 2007)
17 (Nicholls et al., 2001)
18 (Fendinger et al., 1997)
19 (Watanabe et al., 1984a)
20 (Watanabe et al., 1984b)
21 (Jacobs et al., 1987)
22 (Chau et al., 1992)
23 (Fent, 1996)
24 (Fent and Muller, 1991)
25 (Schnaak et al., 1997)
26 (Petrovic and Barcelo, 2000)
27 (Cantero et al., 2004)

Table 4-1. Occurrence of Trace Organic Chemicals Included in This Study (continued).

Chemical	Chemical Class	Use	Max Mean (µg/kg)	Max Median (µg/kg)	Range (µg/kg)	Data Source
High Priority (continued)						
N-MeFOSAA	PFCs and Precursors	Surface Coatings	80	74	31 – 153	9
PFHxS	PFCs and Precursors	Surface Coatings	NA	NA	ND – 110	7.9
PFOS	PFCs and Precursors	Surface Coatings	1,732	1,015	ND – 5,383	8
PFDS	PFCs and Precursors	Surface Coatings	116	49	ND – 426	9
Bisphenol A (BPA)	Plasticizers	Plasticizer	1,430	1,000	100 – 4,600	2
4-Cumylphenol	Surfactants	Detergent Metabolite	2,535	2,535	ND – 5,030	2
4-tert-octyl phenol	Surfactants	Detergent Metabolite	952	937	167 – 2,400	2
Low Priority						
Acetyl Cedrene	PPCPs	Low Concentration	31,300	NA	NA	3
Tonalide (AHTN)	PPCPs	Fragrance material	17,700	4,070	78 – 27,000	2, 3
Azithromycin	PPCPs	Antibiotic	831	278	ND – 5,205	1
BLS	PPCPs	Fluorescent whitening agent	5,400	NA	5,400 – 5,500	13
DAS 1	PPCPs	Fluorescent whitening agent	72,000	NA	86,000 – 112,000	13
Diphenhydramine	PPCPs	Antihistamine	943	424	12 – 7,018	1, 2
Diphenyl Ether	PPCPs	Fragrance material	99,600	NA	NA	3
DSBP	PPCPs	Fluorescent whitening agent	37,000	NA	31,000 – 50,000	12
Galaxolide lactone (HHCB-lactone)	PPCPs	Fragrance material metabolite	1,800	1,600	800 – 3,500	13
Hexyl salicylate	PPCPs	Fragrance material	1,500	NA	NA	3
Hexylcinnamic aldehyde	PPCPs	Fragrance material	4,100	NA	NA	3
Ibuprofen	PPCPs	Analgesic	614	142	ND – 11,900	1
Methyl ionone (gamma)	PPCPs	Fragrance material	3,800	NA	NA	3
Minocycline	PPCPs	Antibiotic	626	430	ND – 8,650	1
Phantolide (AHMI)	PPCPs	Fragrance material	800	700	4 – 843	15
Sulfanilamide	PPCPs	Antibiotic	470	99	191 – 15,600	1
Thiabendazole	PPCPs	Anthelminitic	913	6	1 – 5,000	1, 2
Traseolide (ATII)	PPCPs	Fragrance material	700	700	200 – 1,000	13
17α-Dihydroequilin	Steroidal Chemicals	Steroid hormone	21	20	ND – 98	1
17α-Estradiol	Steroidal Chemicals	Steroid hormone	23	21	ND – 49	1
17β-Estradiol (E2)	Steroidal Chemicals	Steroid hormone	33	22	ND – 355	1
Androstenedione	Steroidal Chemicals	Steroid hormone	324	173	ND – 1,520	1
Androsterone	Steroidal Chemicals	Steroid hormone	124	80	ND – 1,030	1
Equilenin	Steroidal Chemicals	Steroid hormone	16	11	ND – 61	1
Equilin	Steroidal Chemicals	Steroid hormone	35	23	ND – 107	1
Estriol (E3)	Steroidal Chemicals	Steroid hormone	38	25	ND – 232	1
Estrone (E1)	Steroidal Chemicals	Steroid hormone	150	150	ND – 965	1, 2
Etiocholanolone	Steroidal Chemicals	Androgen metabolite	529	NA	NA	16
Iso E super (OTNE)	PPCPs	Fragrance Material	30,700	NA	NA	3
Musk ketone (MK)	PPCPs	Fragrance Material	1,300	NA	NA	3
Norethindrone	Steroidal Chemicals	Synthetic hormone	95	22	ND – 1,360	1
Norgestrel	Steroidal Chemicals	Synthetic hormone	65	42	ND – 1,300	1

NA = Not available; ND = not detected

Data Sources:

1 (U.S. EPA, 2009i)
2 (Kinney et al., 2006)
3 (Difrancesco et al., 2004)
4 (Sellstrom and Jansson, 1995)
5 (Morris et al., 2004)
6 (D'Eon et al., 2009)
7 (Yoo et al., 2009)
8 (Guo et al., 2008)
9 (Higgins et al., 2005)
10 (Bossi et al., 2008)
11 (Sinclair and Kannan, 2006)
12 (Loganathan et al., 2007)
13 (Poiger et al., 1998)
14 (Kupper et al., 2004)
15 (Herren and Berset, 2000)
16 (Tan et al., 2007)
17 (Nicholls et al., 2001)
18 (Fendinger et al., 1997)
19 (Watanabe et al., 1984a)
20 (Watanabe et al., 1984b)
21 (Jacobs et al., 1987)
22 (Chau et al., 1992)
23 (Fent, 1996)
24 (Fent and Muller, 1991)
25 (Schnaak et al., 1997)
26 (Petrovic and Barcelo, 2000)
27 (Cantero et al., 2004)

Table 4-1. Occurrence of Trace Organic Chemicals Included in This Study (continued).

Chemical	Chemical Class	Use	Max Mean (µg/kg)	Max Median (µg/kg)	Range (µg/kg)	Data Source
		Low Priority (continued)				
Progesterone	Steroidal Chemicals	Steroid hormone	327	143	ND – 1,290	1
Testosterone	Steroidal Chemicals	Steroid hormone	162	94	ND – 2,040	1
β-Estradiol-3-benzoate	Steroidal Chemicals	Synthetic hormone	133	23	ND – 1,850	1
N-alkanes (polychlorinated)	Aliphatics	Flame retardant	19,620	11,800	1,800 – 93,100	17
Polydimethylsiloxane (PDMS)	Aliphatics	Organosilicone polymer	821,400	676,000	284,000 – 1,664,000	18
Polyorganosiloxanes	Aliphatics	Organosilicone polymer	8,500	ND	8,310 – 144,000	19, 20
Propene (trichloro)	Aliphatics	Herbicide intermediate	NA	1,140	ND – 167,000	21
Dibutyltin	Organotins	Anti-fouling agent	1,500	NA	ND – 3,400	22, 23
Monobutyltin	Organotins	Heat stabilizer/ anti-fouling agent	780	NA	ND – 3,000	23, 24
Tributyltin	Organotins	Anti-fouling Agent	1,100	3,000	ND – 10,000	22, 23, 25
Hydroquinone	Phenols	Photographic developing/skin whitening agent	NA	2,550	138 – 223,000	21
Cresyldiphenyl phosphate	Phosphate Esters	Plasticizer/flame retardant	NA	18,900	607 – 179,000	21
$C_{10}EO_x$ (Alcohol Ethoxylates)	Surfactants	Surfactant	26,140	NA	ND – 70,000	26
$C_{11}DEA$ (Coconut Diethanol Amide)	Surfactants	Surfactant	2,680	NA	300 – 6,200	26
$C_{12}EO_x$ (Alcohol Ethoxylates)	Surfactants	Surfactant	93,000	NA	1,000 – 122,000	26, 27
$C_{13}DEA$ (Coconut Diethanol Amide)	Surfactants	Surfactant	4,640	NA	200 – 10,500	26
$C_{14}EOx$ (Alcohol Ethoxylates)	Surfactants	Surfactant	44,250	NA	4,500 – 77,000	26
$C_{15}DEA$ (Coconut Diethanol Amide)	Surfactants	Surfactant	2,320	NA	ND – 7,000	26
$C_{16}EOx$ (Alcohol Ethoxylates)	Surfactants	Surfactant	89,500	NA	1,300 – 141,000	26, 27
$C_{17}DEA$ (Coconut Diethanol Amide)	Surfactants	Surfactant	1,460	NA	ND – 5,500	26
$C_{18}EOx$ (Alcohol Ethoxylates)	Surfactants	Surfactant	4,000	NA	ND – 20,000	26
Poly(ethylene glycol)s	Surfactants	Polymer	11,720	NA	1,700 – 31,000	26

NA = Not available; ND = not detected

Data Sources:

1 (U.S. EPA, 2009i)
2 (Kinney et al., 2006)
3 (Difrancesco et al., 2004)
4 (Sellstrom and Jansson, 1995)
5 (Morris et al., 2004)
6 (D'Eon et al., 2009)
7 (Yoo et al., 2009)
8 (Guo et al., 2008)
9 (Higgins et al., 2005)
10 (Bossi et al., 2008)
11 (Sinclair and Kannan, 2006)
12 (Loganathan et al., 2007)
13 (Poiger et al., 1998)
14 (Kupper et al., 2004)
15 (Herren and Berset, 2000)
16 (Tan et al., 2007)
17 (Nicholls et al., 2001)
18 (Fendinger et al., 1997)
19 (Watanabe et al., 1984a)
20 (Watanabe et al., 1984b)
21 (Jacobs et al., 1987)
22 (Chau et al., 1992)
23 (Fent, 1996)
24 (Fent and Muller, 1991)
25 (Schnaak et al., 1997)
26 (Petrovic and Barcelo, 2000)
27 (Cantero et al., 2004)

4.2.2 United States Geological Survey Biosolids Survey

The USGS survey (Kinney et al., 2006) of municipal biosolids destined for land application is of direct relevance to this current study, but the sample size was significantly smaller than that employed in the TNSSS. Biosolids were collected from nine sources, with some of the biosolids being commercially available to consumers. The biosolids originated from utilities in seven states and were analyzed for 87 organic wastewater contaminants. Chemicals included many TOrCs included in the present study, but also many naturally occurring chemicals (i.e., cholesterol) that were excluded from the present study. To enable comparison of the occurrence data from the USGS survey, concentrations were converted to a biosolids dry weight basis from organic-carbon normalized concentrations (as originally reported).

4.3 Literature Surveys of Trace Organic Chemicals in Municipal Sludge

Occurrence data for TOrCs not included in either the USGS survey or the TNSSS were obtained from the primary literature examining the presence of TOrCs in sewage sludge and biosolids. Two recent literature reviews (Harrison et al., 2006; Hydromantis, 2009b) were extremely useful in identifying primary literature sources. To ensure the concentrations reported reflected municipal biosolids as opposed to sludge contaminated from industrial sources, every attempt was made to locate the original studies to verify the source of the sludge and the reported concentrations. When this was possible, the primary sources are cited in Table 4-1.

4.4 Occurrence of Selected TOrCs in Biosolids-Amended Soils

Most studies documenting the presence of the targeted TOrCs in biosolids-amended soils have focused on either the PPCPs or on the brominated flame retardants (BFRs). For example, ciprofloxacin was measured in experimental sludge-amended soils 8 to 21 months post application (at a loading rate of 25 metric tons per hectare) at levels between 270 to 400 $\mu g/kg_{dw}$ (Golet et al., 2002). Norfloxacin was measured in the same samples at levels between 270 and 320 $\mu g/kg_{dw}$. Many of the targeted TOrCs were also detected in biosolids-amended soils by the USGS, though concentrations of some TOrCs (i.e., bisphenol A, triclosan (TCS)) were detected at higher concentrations in the control (minimally affected) site than at the site receiving biosolids (Kinney et al., 2008). Still, concentrations of some of the targeted TOrCs were quite high: hexahydro hexamethylcyclopentabenzopyran (HHCB) was measured in soil as high as 2,770 $\mu g/kg_{dw}$.

Several studies reported the antimicrobial chemicals TCS and triclocarban (TCC) in biosolids-amended soils. Both TCC and TCS were also measured in experimental plots of biosolids-amended soils in the U.S. Ranges of 45-53 $\mu g/kg_{dw}$ and 900-1250 $\mu g/kg_{dw}$ for TCS and TCC, respectively, in the topsoil of plots receiving annual biosolids applications for 33 years were reported. However, the control plot concentrations were quite high, at approximately 20 and 750 $\mu g/kg_{dw}$ for TCS and TCC, respectively (Xia et al., 2010). Another study also observed TCC and TCS in samples collected from biosolids-amended soils from < 1 to 3 years after the last application of biosolids. Concentrations differed depending on the year sampled, but ranged from 1.20 to 65.1 $\mu g/kg_{dw}$ for TCC and <0.05 to 1.02 for TCS $\mu g/kg_{dw}$ (Cha and Cupples, 2009).

The synthetic musks HHCB and acetyl-hexamethyl--tetrahydronaphthalene (AHTN) were detected in amended agricultural fields at concentrations of 2.0 and 2.6 $\mu g/kg_{dw}$, respectively, on the first day after application of biosolids. Similar concentrations were detected two weeks later, but concentrations were nonquantifiable four weeks after application. The ratio of concentration in the biosolids to concentration in the soil (on day one) were approximately 37 and 235 for AHTN and HHCB, respectively, indicating significant dilution effects, as would be predicted (Yang and Metcalfe, 2006).

Biosolids-amended soils in Spain, 1 and 6 years following the last application of biosolids, contained PBDEs; total PBDE (ΣPBDE) concentrations ranged from 30 to 1185 $\mu g/kg_{dw}$ (Ejarrat et al., 2008). The ΣPBDE concentration at the reference site (not amended) was 20.7 $\mu g/kg_{dw}$. Biosolids-amended soils in Sweden sampled 3 to 18 years after the last application of biosolids contained ΣPBDE concentrations of 0.063 and 3900 $\mu g/kg_{dw}$ (Sellstrom et al., 2005). The reference site concentrations (no biosolids amendment) ranged from 0.033 to 1.9 $\mu g/kg_{dw}$. Xia et al. (2010) reported ΣPBDEs concentrations of 120-650 $\mu g/kg_{dw}$ in the topsoil of plots receiving annual biosolids applications for 33 years. The control plot contained a ΣPBDE concentration of approximately 25 $\mu g/kg_{dw}$. In short, PBDEs have been routinely detected in biosolids-amended soils even after considerable time has elapsed. However, relatively low levels have also been detected in control plots, suggesting sources of PBDEs other than biosolids (i.e., atmospheric deposition) may also contribute to PBDE loadings in soils.

4.5 Conclusions

Many TOrCs are present, and at wide ranges of concentrations, in municipal biosolids (Table 4-1). For at least some of the selected TOrCs, there appears to be substantial variability among the reported concentrations. For example, the mean concentration of TCC reported in the TNSSS (Table 4-1) was approximately 40 mg/kg, whereas the median concentration was substantially lower (~ 22 mg/kg). Others have reported TCC concentrations in biosolids in the 5 - 10 ppm range (Cha and Cupples, 2009). While national surveys provide the most robust and extensive data sets, large differences between the mean and median suggest that extra attention must be paid when determining what concentration of a TOrC in biosolids is truly representative of municipal biosolids in the U.S. For some TOrCs, such as the PFCs and PFC precursors, broader surveys (more analytes) will likely be needed to ensure representative concentrations are used in risk assessments. Modeling efforts can predict expected TOrC concentrations, but efforts need to be validated, at least on a limited scale, to ensure accurate and meaningful data are used for risk assessments.

Relatively few studies have documented TOrC presence in biosolids-amended soils. What has been documented suggests high variability with respect to concentrations in both biosolids-amended soils and at control sites. Substantial variability between biosolids-amended soils may result from real variability in the concentrations in the applied biosolids and/or from sampling and spreading errors inherent with large scale land application of biosolids. Several studies have attempted to document the dissipation of TOrCs in biosolids-amended soils based on field sampling (Yang and Metcalfe, 2006; Lozano et al., in press), but efforts were hampered by uncertainties about biosolids application dates and rates and other "real-world" heterogeneity issues. Clearly, additional studies documenting the occurrence and dissipation of the targeted TOrCs in biosolids and biosolids-amended soils are needed.

Table 4-2 provides a summary of the general occurrence data availability for the high priority TOrCs included in this study. This evaluation was limited to the availability of data for TOrCs in sludge and biosolids, and thus does not include an evaluation of the data availability for TOrC occurrence in biosolids-amended soils. The decision criteria used to bin the TOrC classes or subclasses are provided, though in some cases, expert judgment was required to consider data that did not readily fit within the framework developed. Given the selection and prioritization process for inclusion of TOrCs in this study, the general availability of occurrence data for the high priority TOrCs is judged to be quite high.

Table 4-2. Summary of Occurrence Data Availability for the High Priority TOrCs.

Chemical Class	Data Availability
BFRs	Tier 3
PFCs and PFC Precursors	Tier 1
PPCPs: Antimicrobials	Tier 3
PPCPs: Antibiotics	Tier 3
PPCPs: Synthetic Musks	Tier 3
PPCPs: Other	Tier 3
Plasticizers	Tier 3
Steroidal Chemicals	Tier 3
Surfactants	Tier 3

Data Availability Ranking Decision Criteria:

Tier 0 (No Data)	Essentially no data were available of this type for this class or subclass of TOrCs, including data that could be used for modeling.
Tier 1	For the majority of TOrCs in this class or subclass, available data are derived from single peer-reviewed occurrence studies.
Tier 2	For the majority of TOrCs in this class or subclass, available data are derived from multiple peer-reviewed occurrence studies employing appropriate analytical protocols such as isotope dilution mass spectrometry.
Tier 3	For the majority of TOrCs in this class or subclass, available data are derived from large, nationally-representative occurrence studies employing analytical protocols of the highest caliber.

CHAPTER 5.0

MOBILITY OF BIOSOLIDS-BORNE TRACE ORGANIC CHEMICALS IN SOILS

5.1 Introduction

Biosolids-borne trace organic chemicals (TOrCs) applied to soils are subject to transport off-site via processes such as soil erosion and runoff, leaching, and volatilization to non-targeted (e.g., air, water, soil, and biological) media. The mobility of TOrCs in biosolids-amended soils is determined by many factors including physicochemical properties of the chemicals, soil characteristics and properties, environmental conditions such as temperature and rainfall, the mechanism and rate of biosolids application, and the type of biosolids applied. Data regarding the potential mobility of the targeted TOrCs related to transport processes are discussed herein. Studies involving both liquid and dewatered municipal biosolids were considered. Transfer to biological media (e.g., plants and earthworms) is discussed in Chapter 7.0

For the majority of the targeted TOrCs, the dominant process affecting mobility in biosolids-amended soils is sorption. This is inherent to the TOrCs that accumulate in biosolids, since sorption to influent wastewater suspended solids (primary sludge) and activated sludge flocs (secondary sludge) is the major reason for TOrCs in biosolids in the first place. Though volatilization is possible for a few TOrCs, highly volatile chemicals would be expected to volatilize during wastewater conveyance and treatment, and are not expected to accumulate in biosolids. Similarly, highly soluble chemicals are expected to partition primarily into effluent rather than residual solids. Thus, the focus of the majority of this chapter is on the availability of experimentally-derived and modeled physicochemical data, particularly with respect to sorption, that would enable predictions of TOrC mobility in biosolids-amended soils. Available intermediate and field-scale data documenting the potential for leaching or volatilization of the select TOrCs from soil columns and field plots are discussed.

The targeted TOrCs sorb to soil and/or biosolids via mechanisms somewhat different from the sorption mechanism for many of the biosolids-borne TOrCs previously evaluated by U.S. EPA (see Table 2-2). For traditional hydrophobic organic contaminants such as polychlorinated biphenyls (PCBs), the primary mechanism responsible for sorption and retention in soils is hydrophobic portioning to organic matter. For such chemicals, the octanol-water partitioning coefficient (K_{ow}, a general indicator of the chemical hydrophobicity) or the chemical's aqueous solubility (S_w) can be used to predict the organic-carbon normalized soil partition coefficient (K_{oc}) via a variety of linear free energy relationships (Schwarzenbach et al., 2003). The K_{oc} value can then be used to predict the soil-water distribution coefficient (K_d) for a given soil, if the fraction organic carbon (f_{oc}) of the soil is known. Though available S_w values for the targeted TOrCs were tabulated in this study, the use of log K_{ow} values for predicting K_{oc}

values is generally preferred. Hydrophobic partitioning to organic matter (and thus this method for predicting sorption) is likely important for some of the neutral TOrCs targeted in this study (e.g., triclocarban [TCC], polybrominated biphenyl ethers [PBDEs]). However, many of the targeted TOrCs are ionogenic and can exist as charged species depending on the pH of the soil pore water. For charged TOrCs, additional physicochemical parameters both of the TOrC and of the biosolids-amended soil are needed to accurately predict sorption.

Some of the targeted ionogenic TOrCs can exist as anions depending on the pH, while others will exist as cations, and a third group of targeted ionogenic TOrCs can exist as zwitterions (exhibiting both cationic and anionic properties simultaneously). These pH-dependent forms of the TOrCs may all sorb via different mechanisms, complicating any efforts to apply generic hydrophobic partitioning models to predict chemical mobility in soils. For the organic acids (i.e., TOrCs that form anions), the primary sorption mechanism may still be hydrophobic partitioning to organic matter, though the anionic form will generally exhibit much weaker sorption than the neutral form. For these TOrCs, knowledge of the hydrophobicity (i.e., K_{ow}) of the neutral form of the TOrC and the pH at which the charged species becomes dominant (i.e., acid dissociation constant, pK_a) is generally sufficient to estimate the K_d for the TOrC. When using the pK_a to predict a pH-dependent K_d value, it is generally assumed that sorption of the anionic form of the chemical is negligible.

However, some of the targeted TOrCs, such as the perfluorochemicals (PFCs), exhibit significant sorption to organic matter even when present as anions. Given that most PFCs are present as anions at environmentally-relevant pH values, the sorption of the charged species cannot be neglected. Sorption via surface complexation or cation bridging may also be possible for anionic TOrCs that can complex metal cations, such as those associated with carboxylate functional groups. For TOrCs that can exists as cations (i.e., organic bases), sorption of the charged species is often the dominant mechanism, as the positively charged species typically exhibits stronger sorption to negatively charged organic matter and mineral surfaces. Zwitterionic TOrCs can behave both as organic cations and anions, though the stronger sorption of organic cations typically dominates. Thus, the cation exchange capacity (CEC) of the soil is an important parameter for predicting the sorption of organic bases and zwitterions that sorb via cation exchange. For TOrCs that sorb to soils primarily via cation exchange, normalization of the K_d values to the CEC of the soil can also be performed to enable comparisons between soils.

To predict and model the sorption (and thus the mobility) of all TOrCs, but particularly for ionogenic TOrCs, it is important to understand the nature of the sorption process and, especially, the variation in the strength of sorption with chemical concentration. In other words, it is important to know whether a distribution coefficient derived in a laboratory setting at a particular concentration or range of concentrations can be linearly extrapolated to other concentrations more relevant of field conditions. To evaluate sorption linearity in aquatic systems, sorption isotherms are typically constructed using a range of aqueous concentrations. The isotherms are typically evaluated using the Freundlich equation:

$$C_s = K_f (C_w)^n \qquad 5.1\text{-}1$$

where n is an empirically-derived parameter that indicates the linearity of the isotherm ($n = 1$ for a linear isotherm). When Freundlich n values are less than 1, some degree of sorption site limitation and/or variability in the free energies associated with sorption sites can be inferred. Freundlich n values greater than 1 are less common, and may be indicative of self-enhancing sorption (Schwarzenbach et al., 2003). It is also important to note that when $n \neq 1$, the units of K_f are dependent on n, making comparison of K_f values difficult. For this reason, only K_d values and K_{oc} values were tabulated in this chapter for the targeted TOrCs, though n values were provided to give some indication as to the sorption linearity that could be expected in biosolids-amended soils.

5.2 Targeted Physicochemical Data

The availability of physicochemical data useful in modeling sorption was assessed. Specifically, information regarding a TOrC's S_w and log K_{ow} were sought, as well as log K_d, log K_{oc}, Freundlich n values, and pK_a values, if applicable. Values and data sources are identified in Table 5-1. As volatilization is generally not an important release process for most of the selected TOrCs, physicochemical data related to volatilization potential (e.g., vapor pressure, Henry's Law constants) were not tabulated. The potential importance of volatilization in release of TOrCs from biosolids-amended soils is included in Section 5.4 for the relevant TOrCs.

5.3 Sorption of Biosolids-Borne Trace Organic Chemicals in Soils

Significant past work on organic chemical sorption by solid matrices suggests that the traditional organic chemical partitioning paradigm is appropriate for most hydrophobic organic chemicals. This paradigm envisions that equilibrium between the aqueous and solid phases is controlled by the hydrophobicity of the chemical (as measured by its log K_{ow} value) and the fraction of organic carbon in the solid phase (i.e., f_{oc}). As a first approximation, this paradigm will likely hold for many of the selected TOrCs. However, as some of the selected TOrCs are ionogenic, sorption mechanisms other than organic matter partitioning may be very important for specific classes of TOrCs. When relevant, these alternative mechanisms are discussed.

The majority of studies discussed in this chapter do not address desorption of chemicals from solid matrices. Typical transport models assume reversible adsorption of sorbates; however, desorption experiments have shown that release of some of the targeted TOrCs is hysteretic and incomplete (Agyin-Birikorang et al., submitted). Therefore, models that assume complete reversibility of sorption may not be valid for all TOrCs and desorption modeling parameters are needed to better predict mobility in the environment.

5.3.1 Pharmaceuticals and Personal Care Products

Pharmaceuticals and personal care products (PPCPs) included in this study vary in potential mobility in the environment and have been further divided into subclasses to enable meaningful discussions of their mobility behavior. Four subclasses were identified: antimicrobial agents, including triclosan (TCS) and TCC; tetracycline antibiotics, including tetracycline (TC), doxycycline (DTC), and 4-epitetracycline; fluoroquinolone antibiotics, including ciprofloxacin (CIP) and ofloxacin; and synthetic musks, including acetyl-hexamethyl-tetrahydronaphthalene (AHTN) and hexahydro hexamethyl cyclopentabenzopyran (HHCB). No mobility data were found for an additional targeted PPCP, miconazole.

Table 5-1. Physicochemical Parameters for Targeted Trace Organic Chemicals Related to Mobility.

Chemical	Chemical Class	Log S_w (mg/L)	pK_a	Log K_{ow}	Log K_d (L/kg)	Freundlich *n*	Log K_{oc} (L/kg)	Data Sources
Cimetidine	PPCPs				1.04	1.51	2.48	1
Ciprofloxacin (CIP)	PPCPs	0.51 to 4.25	3.01, 6.14, 8.7, 10.58	-1.1 to 1.6	2.4 to 4.3	0.63 to 1.14	3.05 to 4.79	2,3,4,5,6,7,8,9
Doxycycline (DTC)	PPCPs		3.02, 7.97, 9.15	-0.02				3
Galaxolide (HHCB)	PPCPs	0.24 to 0.28		4.6 to 5.9	3.26 to 4.22	1.06 to 1.15	3.63 to 5.96	10,11,12,13,14,15
Ofloxacin	PPCPs		5.97, 8.28	0.35	2.49 to 3.66	0.92 to 1.64	4.61 to 5.51	5,7,16
Tetracycline (TC)	PPCPs	2.36 to 4.75	3.32, 7.7 to 7.78, 9.58 to 9.7	-1.97 to 0.47	1.96 to 5.49	0.49 to 1.61		3,4,17,18,19,20,21,22
Tonalide (AHTN)	PPCPs	0.08 to 0.1		4.84 to 5.8	3.38 to 4.28	0.98 to 1.2	3.68 to 6.04	10,11,12,13,14,15
Triclocarban (TCC)	PPCPs	-1.34 to -0.19	12.7	3.5 to 4.9	2.64 to 3.07	0.50 to 0.99	3.73 to 4.86	1,23,24,25
Triclosan (TCS)	PPCPs	0 to 1	7.9 - 8.1	2.39 to 4.8	2.10 to 2.42	0.87 to 1.57	3.54 to 4.86	1,11,23,24,26,27
17α-ethynylestradiol (EE2)	Steroidal Chemicals	0.49 to 0.68	10.4	4.02 to 4.15	0.37 to 2.85	0.17 to 2.17	2.91 to 4.39	28,29,30,31,32,33,34,35,36,37, 38,39
Mestranol (MeEE2)	Steroidal Chemicals	-0.52	~13	4.67		1.28	4.4	31,32,37
BDE 8/11	BFRs	-1.06		5.03 to 5.83				40
BDE 15	BFRs	-0.89 to -0.1	0.09 to 0.38	5.48 to 5.86				41,42,43,44
BDE 17	BFRs	-1.59		5.52 to 5.88				40
BDE 28	BFRs	-1.15 to -0.48		5.8 to 5.98				41,42,43,44
BDE 46	BFRs	-2.84 to -1.02		6.01 to 6.78			4.95	41,42,43,44,45
BDE 66	BFRs	-1.74		6.73				44
BDE 77	BFRs	-2.22		6.96				44
BDE 85	BFRs	-6.05 to -2.22		6.57 to 7.66	1.95 to 4.65		5.09	40,45
BDE 99	BFRs	-6.05 to -1.41		6.53 to 7.66	1.95 to 4.65		5.09	40,41,42,43,44,45
BDE 100	BFRs	-1.4 to -1.3		6.53 to 7.42	1.95 to 4.65		5.09	43,44,45
BDE 105	BFRs			7.42	1.95 to 4.65		5.09	45
BDE 119	BFRs	-6.05 to -4.1		6.71 to 7.66	1.95 to 4.65		5.09	40,45
BDE 126	BFRs			7.42	1.95 to 4.65		5.09	45

Data Sources:

1 (Barron et al., 2009)
2 (Golet et al., 2003)
3 (Qiang and Adams, 2004)
4 (Thiele-Bruhn, 2003)
5 (Drakopoulos and Ioannou, 1997)
6 (Carrasquillo et al., 2008)
7 (Nowara et al., 1997)
8 (Uslu et al., 2008)
9 (Cordova-Kreylos and Scow, 2007)
10 (Carballa et al., 2008)
11 (Xia et al., 2005)
12 (Litz et al., 2007)
13 (Balk and Ford, 1999a)
14 (Ternes et al., 2004)
15 (Carballa et al., 2007a)
16 (Drillia et al., 2005)
17 (Kim et al., 2005)
18 (Figueroa et al., 2004)
19 (Pils and Laird, 2007)
20 (Sassman and Lee, 2005)
21 (Gu and Karthikeyan, 2005a)
22 (Wang et al., 2008)
23 (Wu et al., 2009)
24 (Ying et al., 2007)
25 (Snyder, 2009)
26 (Aranami and Readman, 2007)
27 (Heidler and Halden, 2007)
28 (Ying and Kookana, 2005)
29 (Ying et al., 2003)
30 (Lee et al., 2007)
31 (Ying et al., 2002)
32 (Shareef et al., 2006)
33 (Hildebrand et al., 2006)
34 (Bonin and Simpson, 2007)
35 (Yu et al., 2004)
36 (Yu and Huang, 2005)
37 (Lai et al., 2000)
38 (Stumpe and Marschner, 2009)
39 (Lee et al., 2003)
40 (Palm et al., 2002)
41 (Gouin and Harner, 2003)
42 (Kuramochi et al., 2007)
43 (Wania and Dugani, 2003)
44 (Tittlemier et al., 2002)
45 (Litz, 2002)
46 (Wania and Dugani, 2003)
47 (Sun et al., 2008b)
48 (Sun et al., 2008c)
49 (Arp et al., 2006)
50 (Carmosini and Lee, 2008a)
51 (Liu and Lee, 2007)
52 (Liu and Lee, 2005)
53 (Higgins and Luthy, 2006)
54 (Goss, 2008)
55 (Burns et al., 2008)
56 (Johnson et al., 2007)
57 (3M, 2000)
58 (Tan et al., 2007)
59 (Tan et al., 2008)
60 (Pan et al., 2009)
61 (Navarro et al., 2009)
62 (Zhou, 2006)
63 (Ahel and Giger, 1993)
64 (Yamamoto et al., 2003)
65 (Johnson et al., 1998)

Table 5-1. Physicochemical Parameters for Targeted Trace Organic Chemicals Related to Mobility (continued).

Chemical	Chemical Class	Log S_w (mg/L)	pK_a	Log K_{ow}	Log K_d (L/kg)	Freundlich *n*	Log K_{oc} (L/kg)	Data Sources
BDE 137	BFRs			7.91				44
BDE 152	BFRs	-6.06 to -1.78		7.08 to 8.05				40,41,42,43,44
BDE 153	BFRs	-6.06		7.39				44
BDE 171	BFRs	-2.20		7.49				46
BDE 180	BFRs	-2.20		7.49				46
BDE 181	BFRs	-6.67 to -2.2		7.49 to 9.44				40,46
BDE 183	BFRs	-2.82 to -2.2		7.49				46,44
BDE 184	BFRs	-2.20		7.49				46
BDE 190	BFRs	-6.67 to -2.2		7.49 to 9.44				40,46,44
BDE 191	BFRs	-2.20		7.49				46
BDE 196	BFRs	-2.64		7.9				46
BDE 197	BFRs	-7.95 to -2.64		7.9 to 10.33				40,46
BDE 201	BFRs	-2.64		7.9				46
BDE 203	BFRs	-2.64		7.9				46
BDE 206	BFRs	-3.10		8.3				46
BDE 207	BFRs	-3.10		8.3				46
BDE 208	BFRs	-3.10		8.3				46
BDE 209	BFRs	-10.89 to -3.55		8.7 to 11.15				40,46
TBBPA	BFRs	0.66	7.5, 8.5	0.39 to 5.34	1.27 to 2.60	0.99		11,47,48
4:2 FTOH	PFCs and Precursors	2.99		3.28 to 3.3	-0.15		0.93	49,50,51
6:2 FTOH	PFCs and Precursors	1.27		4.54 to 4.70	0.29 to 1.23		2.43	49,50,51
8:2 FTOH	PFCs and Precursors	-0.71		5.58 to 6.14	0.58		4.13	49,50,52
10:2 FTOH	PFCs and Precursors	-2.22 to -0.05		7.57	0.79		6.20	49,50,51
PFOA	PFCs and Precursors		2.3 to 3.8	4.3			2.06 to 2.11	49,53,54,55
PFNA	PFCs and Precursors			3.86 to 7.27			2.39 to 2.5	49,53
PFDA	PFCs and Precursors			8.2			2.76 to 2.92	53
PFUnDA	PFCs and Precursors						3.3 to 3.47	53

Data Sources:

1 (Barron et al., 2009)
2 (Golet et al., 2003)
3 (Qiang and Adams, 2004)
4 (Thiele-Bruhn, 2003)
5 (Drakopoulos and Ioannou, 1997)
6 (Carrasquillo et al., 2008)
7 (Nowara et al., 1997)
8 (Uslu et al., 2008)
9 (Cordova-Kreylos and Scow, 2007)
10 (Carballa et al., 2008)
11 (Xia et al., 2005)
12 (Litz et al., 2007)
13 (Balk and Ford, 1999a)
14 (Ternes et al., 2004)
15 (Carballa et al., 2007a)
16 (Drillia et al., 2005)
17 (Kim et al., 2005)
18 (Figueroa et al., 2004)
19 (Pils and Laird, 2007)
20 (Sassman and Lee, 2005)
21 (Gu and Karthikeyan, 2005a)
22 (Wang et al., 2008)
23 (Wu et al., 2009)
24 (Ying et al., 2007)
25 (Snyder, 2009)
26 (Aranami and Readman, 2007)
27 (Heidler and Halden, 2007)
28 (Ying and Kookana, 2005)
29 (Ying et al., 2003)
30 (Lee et al., 2007)
31 (Ying et al., 2002)
32 (Shareef et al., 2006)
33 (Hildebrand et al., 2006)
34 (Bonin and Simpson, 2007)
35 (Yu et al., 2004)
36 (Yu and Huang, 2005)
37 (Lai et al., 2000)
38 (Stumpe and Marschner, 2009)
39 (Lee et al., 2003)
40 (Palm et al., 2002)
41 (Gouin and Harner, 2003)
42 (Kuramochi et al., 2007)
43 (Wania and Dugani, 2003)
44 (Tittlemier et al., 2002)
45 (Litz, 2002)
46 (Wania and Dugani, 2003)
47 (Sun et al., 2008b)
48 (Sun et al., 2008c)
49 (Arp et al., 2006)
50 (Carmosini and Lee, 2008a)
51 (Liu and Lee, 2007)
52 (Liu and Lee, 2005)
53 (Higgins and Luthy, 2006)
54 (Goss, 2008)
55 (Burns et al., 2008)
56 (Johnson et al., 2007)
57 (3M, 2000)
58 (Tan et al., 2007)
59 (Tan et al., 2008)
60 (Pan et al., 2009)
61 (Navarro et al., 2009)
62 (Zhou, 2006)
63 (Ahel and Giger, 1993)
64 (Yamamoto et al., 2003)
65 (Johnson et al., 1998)

Table 5-1. Physicochemical Parameters for Targeted Trace Organic Chemicals Related to Mobility (continued).

Chemical	Chemical Class	Log S_w (mg/L)	pK_a	Log K_{ow}	Log K_d (L/kg)	Freundlich *n*	Log K_{oc} (L/kg)	Data Sources
PFDoDA	PFCs and Precursors			4.09 to 8.23				49
FOSA	PFCs and Precursors			3.21 to 7.58				49
N-EtFOSAA	PFCs and Precursors						3.23 to 3.49	53
N-MeFOSAA	PFCs and Precursors						3.11 to 3.35	53
PFHxS	PFCs and Precursors				0.45 to 0.95		2.4 to 3.1	56
PFOS	PFCs and Precursors			6.3	0.45 to 2.08		2.4 to 2.68	53,56,57
PFDS	PFCs and Precursors			8.2			3.53 to 3.66	53
Bisphenol A (BPA)	Plasticizers	2.08 to 2.48	9.59 to 10.2	3.3 to 3.6	-2.43 to 1.64	1.18	2.89 to 2.98	28,29,58,59,60
4-Cumylphenol	Surfactants			4.1				58,59
4-tert-Octyl phenol	Surfactants	1.10	10.24	4.12 to 5.85	0.78 to 3.25	0.69 - 1.22	3.60 to 4.70	28,29,58,59,61,62,63,64,65

Data Sources:

1	(Barron et al., 2009)	14	(Ternes et al., 2004)	27	(Heidler and Halden, 2007)	40	(Palm et al., 2002)	53	(Higgins and Luthy, 2006)
2	(Golet et al., 2003)	15	(Carballa et al., 2007a)	28	(Ying and Kookana, 2005)	41	(Gouin and Harner, 2003)	54	(Goss, 2008)
3	(Qiang and Adams, 2004)	16	(Drillia et al., 2005)	29	(Ying et al., 2003)	42	(Kuramochi et al., 2007)	55	(Burns et al., 2008)
4	(Thiele-Bruhn, 2003)	17	(Kim et al., 2005)	30	(Lee et al., 2007)	43	(Wania and Dugani, 2003)	56	(Johnson et al., 2007)
5	(Drakopoulos and Ioannou, 1997)	18	(Figueroa et al., 2004)	31	(Ying et al., 2002)	44	(Tittlemier et al., 2002)	57	(3M, 2000)
6	(Carrasquillo et al., 2008)	19	(Pils and Laird, 2007)	32	(Shareef et al., 2006)	45	(Litz, 2002)	58	(Tan et al., 2007)
7	(Nowara et al., 1997)	20	(Sassman and Lee, 2005)	33	(Hildebrand et al., 2006)	46	(Wania and Dugani, 2003)	59	(Tan et al., 2008)
8	(Uslu et al., 2008)	21	(Gu and Karthikeyan, 2005a)	34	(Bonin and Simpson, 2007)	47	(Sun et al., 2008b)	60	(Pan et al., 2009)
9	(Cordova-Kreylos and Scow, 2007)	22	(Wang et al., 2008)	35	(Yu et al., 2004)	48	(Sun et al., 2008c)	61	(Navarro et al., 2009)
10	(Carballa et al., 2008)	23	(Wu et al., 2009)	36	(Yu and Huang, 2005)	49	(Arp et al., 2006)	62	(Zhou, 2006)
11	(Xia et al., 2005)	24	(Ying et al., 2007)	37	(Lai et al., 2000)	50	(Carmosini and Lee, 2008a)	63	(Ahel and Giger, 1993)
12	(Litz et al., 2007)	25	(Snyder, 2009)	38	(Stumpe and Marschner, 2009)	51	(Liu and Lee, 2007)	64	(Yamamoto et al., 2003)
13	(Balk and Ford, 1999a)	26	(Aranami and Readman, 2007)	39	(Lee et al., 2003)	52	(Liu and Lee, 2005)	65	(Johnson et al., 1998)

5.3.1.1 Antimicrobial Agents

TCS and TCC are antimicrobial agents commonly found in biosolids at concentrations in the low tens of parts per million and readily accumulate in sewage sludge (Heidler and Halden, 2007). Both TCC and TCS are generally expected to sorb to solid phases according to the traditional hydrophobic partitioning paradigm, though the pK_a for TCS (~8) suggests significantly less sorption to organic matter at high pH values. Reported log K_{ow} values for TCS are 2.39 - 4.8 (Xia et al., 2005) and reported log K_{ow} values for TCC are 3.5 - 4.9 (Ying and Kookana, 2007; Ying et al., 2007; Snyder, 2009). These values suggest a tendency for TCC and TCS partitioning to soil or sediment that contains significant quantities of organic carbon (OC; Ying and Kookana, 2007; Ying et al., 2007).

Sorption Data

Published studies of the sorption of TCS and TCC are limited. Two studies reported K_d values for TCS and TCC (Barron et al., 2009; Wu et al., 2009). The first study (Barron et al., 2009) conducted batch sorption experiments with TCS and TCC in an agricultural soil and digested wastewater treatment plant (WWTP) sludge. Log K_d values measured for TCS and TCC in the soil were 2.10 and 2.64, respectively. However, considerable nonlinearity was also observed (Barron et al., 2009). The second study conducted batch sorption experiments with TCS and TCC in a silty clay and a sandy loam (Wu et al., 2009). Log K_d values measured in this study were 2.25 - 2.42 for TCS and 2.88 - 3.07 for TCC. Isotherms were nearly linear in these soils. Sorption of both TCS and TCC was greater in the sandy loam despite greater f_{oc} in the silty clay. The authors provided two potential causes. First, this could be due to differences in the type of organic matter in the two soils. Second, the higher clay content in the silty clay may inhibit sorption interactions with the soil organic matter (Wu et al., 2009).

In general, K_d values indicate that TCC has a greater affinity for solid phases than TCS. Sorption of both chemicals can be affected by factors such as f_{oc} and co-solutes, and as discussed above, pH is expected to affect the sorption of TCS more than TCC. The effects of amending soils with biosolids on TCC and TCS was also examined by Wu et al. (2009). Sorption of both TCC and TCS increased in soils after biosolids-amendment. Amending soils with biosolids increased soil f_{oc} and pH, There was no change in TCC sorption over the pH range of 3 - 9 (Wu et al., 2009), but sorption of TCS decreased as solution pH increased over the same range. In biosolids-amended soils, sorption of these chemicals increased despite increasing pH, suggesting that the impact of increasing f_{oc} is a more important effect (Wu et al., 2009).

Sorption Modeling

Quantitative structure activity relationship (QSAR) analyses have been used to calculate K_{ow} and K_{oc} values for both TCS and TCC (Ying and Kookana, 2007; Ying et al., 2007). Results of these calculations were used to estimate partitioning of TCS and TCC in four compartments including air, water, soil, and sediment. More than 80% of the mass of both chemicals was predicted to partition to soil and sediment. As expected, the fraction of TCS predicted for the aqueous phases was greater than that of TCC (Ying and Kookana, 2007; Ying et al., 2007). These results are in general agreement with both the sorption data and the studies of TCS in tile drainage (Lapen et al., 2008b; Edwards et al., 2009).

5.3.1.2 Tetracycline Antibiotics

Unmetabolized tetracycline antibiotics released to WWTPs have the potential to partition to sewage sludge during wastewater treatment where they may ultimately enter the environment through biosolids application. Many identified studies focused only on oxytetracycline (OTC), as opposed to TC. Due to structural similarities and similar pK_a values between OTC, DTC, and TC, in the absence of chemical-specific data, data on the mobility of OTC can suggest the likely mobility of the tetracyclines included in this study (TC, DTC, and 4-epitetracycline).

The chemical structure of TC is shown in Figure 5-1. Tetracycline is an ionogenic chemical with three ionizable functional groups that leads to complex sorption behavior (Sassman and Lee, 2005; Pils and Laird, 2007). An extensive review of the sorptive mechanisms for the tetracycline antibiotics can be found in Carmosini and Lee (2008b). The three functional groups on TC are the tricarbonyl methane system, phenolic diketone moiety, and a dimethylammonium cation. For TC, the pK_{a1}, pK_{a2}, and pK_{a3} values, measured through potentiometric titration, were determined to be 3.32, 7.78, and 9.58, respectively (Qiang and Adams, 2004). There are four possible forms of TC, depending on pH (Sithole and Guy, 1987a), but the cationic form of the tetracyclines dominates sorption, even under pH conditions where the zwitterions dominates (Carmosini and Lee, 2008b and references therein). Cation exchange and surface complexation are the dominant sorption mechanisms (Sithole and Guy, 1987a; Sithole and Guy, 1987b; Figueroa et al., 2004; Kulshrestha et al., 2004; Sassman and Lee, 2005; Carmosini and Lee, 2008b), though there is some disagreement as to whether this is true when the zwitterions predominates (Sithole and Guy, 1987a; Sithole and Guy, 1987b). Models incorporating the mechanisms of sorption have proven accurate at a wide range of pH values (Sassman and Lee, 2005). As a result of the importance of these alternative sorption mechanisms, data on the hydrophobicity (i.e., log K_{ow}) of the tetracyclines is of limited utility, as hydrophobic organic matter partitioning is not the dominant sorption mechanism.

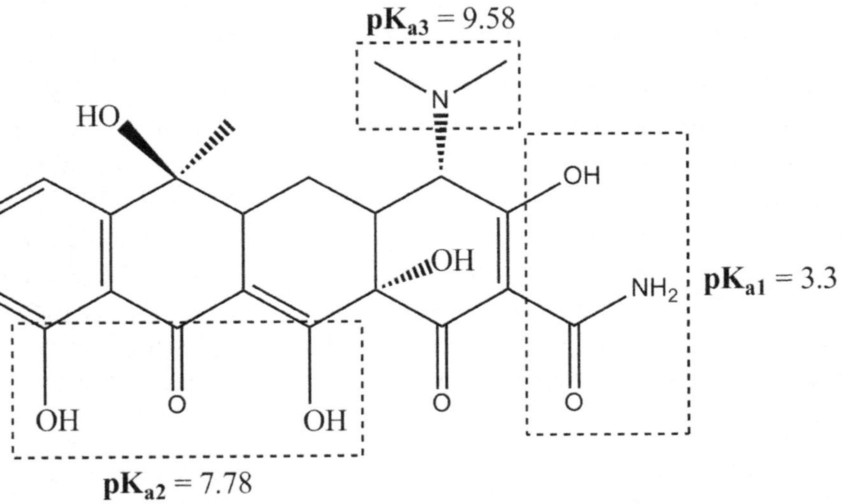

Figure 5-1. Chemical Structure of Tetracycline (TC) with pK_a Values for each of the Three Functional Groups.

Sorption Data

Sorption of tetracycline antibiotics is generally extensive. As much as 95-99% of tetracyclines were sorbed, regardless of soil type (Loke et al., 2002). Furthermore, tetracyclines are capable of sorbing to a wide range of sorbents including humic acids, proteins, organic matter, soil, and oxides. Reported log K_d values for TC range from 1.96 to 5.49 for a variety of sorbents (Figueroa et al., 2004; Sassman and Lee, 2005; Pils and Laird, 2007; Bao et al., 2009). Likely as a result of the multiple sorption mechanisms for the tetracyclines, substantial sorption nonlinearity is commonly observed (Sassman and Lee, 2005; Gu et al., 2007; Wang et al., 2008).

The interaction of TC with organic matter has been the subject of several studies (Sithole and Guy, 1987a; Sithole and Guy, 1987b; Kulshrestha et al., 2004; MacKay and Canterbury, 2005; Pils and Laird, 2007; Bao et al., 2009). Research clearly shows there is an interaction between tetracyclines and organic matter, though as organic matter is usually a significant contributor to the CEC of soils, this would be expected. During sorption experiments, the lowest sorption was observed to organic matter with the lowest ability to complex metals. Addition of Al or Fe to the organic matter, which would lead to a higher concentration of potential bridging cations, increased sorption (MacKay and Canterbury, 2005). Dissolved organic matter (DOM) may also impact tetracycline antibiotic sorption. At low DOM concentrations (1.0 mg/L), OTC sorption increased, whereas sorption decreased at higher concentrations (10 mg/L) of DOM (Kulshrestha et al., 2004). Tetracycline binding to DOM may prevent subsequent surface interactions (Kulshrestha et al., 2004). Interestingly, studies of tetracycline sorption on clay, humic substances, and clay-humic complexes showed that sorption of TC on clay-humic complexes was less than that observed on either clay or the humic substances alone. X-ray diffraction analysis revealed the presence of humic substances in clay interlayers that may block TC access to sorption sites in the clay interlayers (Pils and Laird, 2007).

Effect of pH and Solution Chemistry on Sorption

Sorption of tetracycline antibiotics to solid phases generally decreases as pH increases (Sithole and Guy, 1987a; Sithole and Guy, 1987b; Figueroa et al., 2004; Pils and Laird, 2007). This pH-dependent behavior is expected based on the pK_a values of tetracycline antibiotics which lead to changing charge on the functional groups as pH changes. Additionally, changes in pH can alter the charge of surface particles which would also affect sorption. For instance, from pH 4 to pH 7, the negative charge of both TC and organic matter increases which could lead to increased repulsion between the two (Sithole and Guy, 1987a; Sithole and Guy, 1987b; Pils and Laird, 2007). Additionally, charge on edge sites of clays would become anionic with increasing pH which could also lead to repulsion between these particles and tetracyclines (Sithole and Guy, 1987a; Sithole and Guy, 1987b; Pils and Laird, 2007). Others also note that increasing pH decreases the ability of tetracycline antibiotics to sorb through cation exchange (Sassman and Lee, 2005). However, the decrease in sorption with increasing pH is not as much as would be expected if only cation exchange were contributing to sorption, indicating that the increase in anionic charge of TC with an increase in pH also contribute to reduced sorption (Figueroa et al., 2004). The latter also supports the observation of other sorption mechanisms and the increase in importance of mechanisms such as cation bridging as the zwitterion becomes more dominant.

Increases in solution ionic strength decreases sorption of tetracycline antibiotics in clays (Sithole and Guy, 1987a; Sithole and Guy, 1987b; Figueroa et al., 2004). This trend was also observed in sorption of tetracycline to soil organic matter (Sithole and Guy, 1987a; Sithole and Guy, 1987b). One study used an empirical model to show that the effect is more important for the cationic species of TC than for the zwitterion (Figueroa et al., 2004). Thus, ionic strength effects on the sorption of these TOrCs depends on the species of tetracycline antibiotics present, which in turn is determined by pH.

Sorption of TC in can also vary with co-solutes. Thus, clays saturated with divalent cations sorbed more TC than clays saturated with monovalent cations (Sithole and Guy, 1987a; Sithole and Guy, 1987b; Figueroa et al., 2004; Pils and Laird, 2007). Divalent cations can bridge between negatively charged moieties on TC and negative charges on the surface of clay particles. The interaction take places in the interlayers (Sithole and Guy, 1987a; Sithole and Guy, 1987b; Pils and Laird, 2007). Interaction of TCs with divalent metals can increase or decrease sorption depending on whether the metals are surface bound. Several studies have noted an increase in TC sorption in the presence of divalent metals (Loke et al., 2002; Wang et al., 2008). However, others reported that the presence of divalent metals decreased TC sorption and increased metal solubility (Loke et al., 2002; Gu and Karthikeyan, 2005a).

5.3.1.3 Fluoroquinolone Antibiotics

As with the tetracyclines, sorption of fluoroquinolones to soils does not follow the traditional organic matter partitioning paradigm. Fluoroquinolones are soluble, hydrophilic molecules, and yet several authors point to low potential for fluoroquinolone antibiotics mobility in the environment (Nowara et al., 1997; Uslu et al., 2008). The strong sorption of fluoroquinolones to soils is generally thought to result from cation exchange, surface complexation, and cation bridging.

The chemical structure the fluoroquinolones antibiotic CIP is shown in Figure 5-2, (the structure of ofloxacin is similar). Similar to TC, CIP is an amphoteric molecule having both acidic and basic functional groups. The carboxylic acid and amine groups have pK_a values of 6.14 and 8.17, respectively (Qiang and Adams, 2004), while the corresponding pK_a values for ofloxacin were estimated from spectrofluorimetric data to be 5.97 and 8.28, respectively (Drakopoulos and Ioannou, 1997; Qiang and Adams, 2004). Three forms of the fluoroquinolones are possible at environmentally relevant pH values (Gu and Karthikeyan, 2005b), which suggests that pH-dependent sorption is likely.

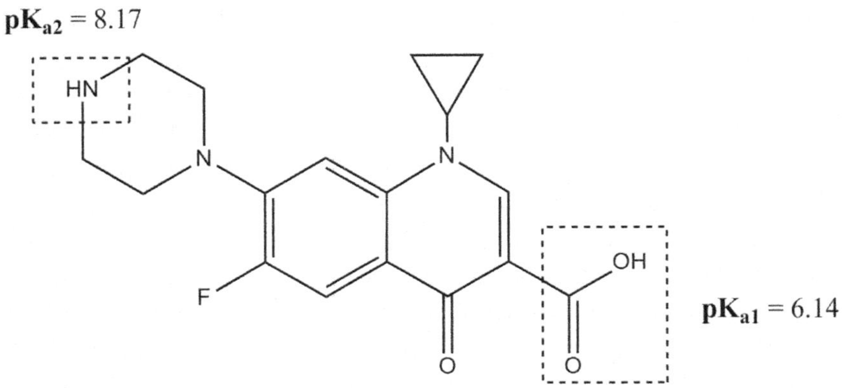

Figure 5-2. Chemical Structure of Ciprofloxacin (CIP) with pK_a Values for each of the Two Functional Groups.

With respect to sorption mechanisms, two studies employing principal component analysis found that CIP sorption is highly correlated with CEC (Carrasquillo et al., 2008; Vasudevan et al., 2009). The primary functional group involved in this cation exchange is the positively charged amine group (Vasudevan et al., 2009). Thus, cation exchange is expected to dominate at lower pH values (<5.5) when the amine group still carries a positive charge (Vasudevan et al., 2009). Principal component analysis also found strong correlation of CIP sorption with complexation involving Fe- and Al- oxides (Carrasquillo et al., 2008; Vasudevan et al., 2009). Complexation likely involves the interaction of soil metal oxides with the CIP carboxyl group (Vasudevan et al., 2009), which is expected to be more prevalent at pH 5.5 - 8. Several studies reported the ability of CIP to form strong complexes with Al- and Fe- oxides (Gu and Karthikeyan, 2005b; Zhang and Huang, 2007; Mackay and Seremet, 2008).

The sorption mechanism that plays a primary role in overall sorption of fluoroquinolones depends on the soil characteristics. Probe compounds with only one functional group have been used to quantify the contribution of these two mechanisms (Mackay and Seremet, 2008). Results of this work indicated that cation exchange would contribute increasingly to overall sorption with increasing CEC of soil. Similarly, complexation with Al- and Fe- oxides contributes increasingly to sorption with increasing Al- and Fe- oxide content of soil. Similar results were obtained in sorption experiments conducted in soil with high CEC over a range of pH values. Sorption decreased at all pH values, even at pH values where the contribution of surface complexation was expected to increase. These results indicate that even increased sorption by surface complexation does not offset the decrease due to less CEC, provided the number of cation exchange sites are not limited (Vasudevan et al., 2009). Surface complexation plays a more significant role in soils where cation exchange sites were limited.

Sorption Data

As with the tetracyclines, sorption of fluoroquinolones is high despite low K_{ow} values. Log K_d values reported for CIP in the literature are 2.4 - 4.3 for a variety of sorbents (Nowara et al., 1997; Cordova-Kreylos and Scow, 2007; Carrasquillo et al., 2008; Uslu et al., 2008; Wu et al., 2008). Log K_d values reported for ofloxacin are 2.49 - 3.66 (Nowara et al., 1997; Drillia et al., 2005).

Several authors noted a lack of a clear relationship between K_d and K_{oc} for fluoroquinolone antibiotics (Tolls, 2001; Drillia et al., 2005; Vasudevan et al., 2009). As organic matter also contributes to a soil's CEC, it is possible that organic matter effects on sorption are accounted for when studying the effects of CEC (Vasudevan et al., 2009). Acknowledging the role of organic matter in cation exchange, Carmosini and Lee (2009) examined the sorption of CIP to dissolved organic matter (DOM), including DOM derived from both digested and undigested biosolids. Using the moderately high K_{doc} values measured, the authors concluded that DOM-facilitated transport could increase the mobility of CIP by up to 15% in biosolids-amended soils.

Effect of pH and Solution Chemistry on Sorption

Fluoroquinolones One study conducted sorption experiments with CIP in 30 soils at pH levels from 3 - 8 (Vasudevan et al., 2009). Sorption trends with varying pH fell into two categories. In the majority of soils, sorption was highest at pH 5.5 and significantly decreased between pH 7 and 8. In a second group of soils, sorption decreased as pH increased over the entire pH range. In general, K_d values were higher in the second group. The first trend represents soils rich in metal oxides, with low CEC, where contributions of surface complexation to sorption are significant. The second trend represents soils in which cation exchange is the dominant sorption mechanism and where contributions of surface complexation were not discernable (Vasudevan et al., 2009). Two additional studies have found CIP sorption results similar to the first sorption trend described above. In both cases, the trend was observed in soils rich in metal oxides (Gu and Karthikeyan, 2005b; Zhang and Huang, 2007). Clearly, the pH-dependent behavior of fluoroquinolone sorption depends on soil characteristics, which in turn determine the dominant sorption mechanisms.

Ionic strength may also play a role in fluoroquinolones sorption, and though sorption of CIP to Al- and Fe- oxides did not depend on ionic strength (Gu and Karthikeyan, 2005b), others have observed differing effects of ionic strength depending on solution pH (Carmosini and Lee, 2009). In this latter study, at low pH, sorption of CIP to DOM decreased with increasing levels of K^+, whereas this effect was not observed at high pH. These data are consistent with a cation exchange mechanism for CIP at low pH (Carmosini and Lee, 2009).

5.3.1.4 Synthetic Musks

As with the antimicrobials TCC and TCS, the synthetic musks are generally expected to sorb according to the traditional hydrophobic partitioning paradigm. Log K_{ow} values for AHTN and HHCB are relatively high: reported AHTN log K_{ow} values range from 4.84 - 5.8, while reported HHCB log K_{ow} values are in the 4.6 - 5.9 range (Balk and Ford, 1999a; Carballa et al., 2007b; Carballa et al., 2008). These K_{ow} values indicate that hydrophobic interactions likely play an important role for AHTN and HHCB mobility in the environment.

Sorption Data

Few studies were available of the sorption of AHTN and HHCB in soils. However, literature is available documenting sorption of these chemicals in wastewater sludge. Log K_d values of 3.38 - 4.28 (AHTN) and 3.26 - 4.22 (HHCB) have been reported for secondary and digested sludge (Ternes et al., 2004; Carballa et al., 2008), indicating strong affinity of the chemicals for the solid phase. One study was identified that conducted sorption experiments in soil. Sorption was essentially linear for HHCB and AHTN (Litz et al., 2007), allowing log K_f values to be interpreted as log K_d values. The reported values were 1.98 - 2.58 for HHCB and 2.18 - 2.82 for AHTN (Litz et al., 2007). For both AHTN and HHCB, stronger sorption was exhibited in soils with higher f_{oc} (Litz et al., 2007).

5.3.1.5 Cimetidine

Cimetidine is the only histamine H_2-receptor antagonist included in this study. One study included this compound in batch sorption experiments, but no information outside of sorption coefficients in an agricultural soil was discussed in the study. A log K_d value of 1.04 was reported, though the study also reported high isotherm nonlinearity (Barron et al., 2009).

5.3.2 Steroidal Chemicals

The only two high priority steroidal chemicals selected for data gap analysis were 17α-ethinyl estradiol (EE2) and mestranol (MeEE2). The reported log K_{ow} values for EE2 are in the 4.02 - 4.15 range (Ying et al., 2002; Lee et al., 2007), while the reported value for MeEE2 is slightly higher (4.67; Ying et al., 2002). These data suggest that the hydrophobic partitioning paradigm should be applicable, and that EE2 sorbs less than MeEE2. In fact, there is good agreement in the literature that hydrophobic partitioning plays an important role in sorption of steroidal chemicals. Only one report identified a lack of correlation between K_{ow} and K_{oc} and the authors noted that the results should be verified due to the fluorescence quenching method used to verify sorption (Yamamoto et al., 2003). One report identified hydrophobic partitioning as the dominant sorption mechanism for EE2 based on consistency of log K_{oc} values measured in three different soils (Lee et al., 2003). However, other authors note that steroidal chemicals sorb to sorbents such as iron oxide that have no significant organic carbon content, suggesting that ion exchange with surface hydroxyl groups also plays a role (Lai et al., 2000).

Sorption Data

Many sorption studies have been completed for EE2 in soils, sludge, minerals, and sediment (Lai et al., 2000; Ying et al., 2002; Ying and Kookana, 2005; Carballa et al., 2008), while relatively few studies were found for MeEE2 (Lai et al., 2000; Shareef et al., 2006). Most studies reported varying degrees of nonlinear sorption of both EE2 and MeEE2 in all solid media (Lee et al., 2003; Hildebrand et al., 2006; Bonin and Simpson, 2007; Stumpe and Marschner, 2009). Sorption was essentially linear in some soils, and log K_d sorption values ranged from 0.37

- 2.09 (Lee et al., 2003; Ying and Kookana, 2005). Sorption of EE2 in soil increased with increasing f_{oc} (Ying and Kookana, 2005; Hildebrand et al., 2006; Bonin and Simpson, 2007).

Sorption in sediment was less than in soils and sorption linearity was considerably more variable (Lai et al., 2000; Ying et al., 2003; Yu et al., 2004; Yu and Huang, 2005). As with sorption in soils, some EE2 sorption to sediment was reported as linear, with log K_d values ranging from 1.24 - 1.99 (Yu and Huang, 2005). The same study reported increasing sorption with increasing concentration of EE2 (Yu and Huang, 2005). Sorption of MeEE2 was somewhat less linear in sediment (Lai et al., 2000). Sorption of both EE2 and MeEE2 in sediment increased with increasing f_{oc} (Lai et al., 2000), though in another study, organic carbon normalized sorption coefficients varied by more than an order of magnitude (Bonin and Simpson, 2007). Sorption of EE2 and MeEE2 to iron oxide, which has no appreciable organic carbon content, has been observed (Lai et al., 2000).

Sorption experiments were also completed with EE2 and clay minerals including kaolinite, goethite, and montmorillonite. For these minerals, EE2 sorption was highly nonlinear (Bonin and Simpson, 2007). One study noted sorption of MeEE2 to clay minerals was greater than that of EE2; however, sorption coefficients were not reported (Shareef et al., 2006). Two studies observed higher EE2 sorption in montmorillonite than in non-expanding clays kaolinite and goethite (Shareef et al., 2006; Bonin and Simpson, 2007). This trend is likely due to sorption of EE2 into the expanding clay interlayer of montmorillonite (Shareef et al., 2006; Bonin and Simpson, 2007). Sorption of EE2 and MeEE2 to montmorillonite was pH dependent (Shareef et al., 2006).

Varying degrees of EE2 desorption have been reported from soil and clay minerals (Hildebrand et al., 2006; Shareef et al., 2006). Desorption of EE2 was greater in soils with lower f_{oc} and clay content. Nearly 100% of EE2 desorbed from sand and silt loam at initial EE2 concentrations of less than 100 nanograms/milliliter (ng/mL), whereas less than 20% desorbed from clay loam and silty clay at the same EE2 concentrations (Hildebrand et al., 2006). Eighty percent of EE2 desorbed from kaolinite and goethite. Less EE2 desorbed from montmorillonite, and desorption varied with pH. Some desorption from montmorillonite was observed at pH 4, but little or no desorption was observed at pH 10. These pH values were selected because they were the original pH of the suspensions (Shareef et al., 2006).

Effects of pH on Sorption

Sorption of EE2 and MeEE2 increases with increasing f_{oc}, but organic carbon is not the only factor controlling sorption of these chemicals. The pK_a values of EE2 and MeEE2 are 10.4 and ~13, respectively (Shareef et al., 2006), so both chemicals will be in their neutral forms at environmentally relevant pH values. If pH-dependent behavior is observed, it is expected to be the result of the pH-dependent behavior of the sorbent. For example, in the sorption of EE2 and MeEE2 onto kaolinite and goethite, sorption was pH-invariant. However sorption of the chemicals to montmorillonite increased with increasing pH (Shareef et al., 2006). Because the surface charge of all three minerals would change with changing pH and because pH-dependent behavior was observed only with montmorillonite, it is not likely that a change in surface charge is the cause of this trend (Shareef et al., 2006). The authors hypothesized that a change in the nature of the flocculation of montmorillonite caused the trend. At high pH values, montmorillonite is arranged in a manner that would allow better access of EE2 and MeEE2 to the mineral's interlayer (Shareef et al., 2006). Desorption also decreased with increasing pH (Shareef et al., 2006). As sorption of EE2 to montmorillonite likely involves diffusion into the clay interlayer (which may be more accessible at pH 10), and sorption into the interlayer is not as reversible a process as is sorption to the external surface, this likely explains why desorption of EE2 decreased with increasing pH (Shareef et al., 2006).

Effects of Co-Solutes on Sorption

Sorption of EE2 and MeEE2 in the environment is likely to occur in the presence of other chemicals. The impacts of estradiol (E2), estrone (E1), naphthalene, phenanthrene, and estradiol valerate as co-solutes on the sorption of EE2 and, in some cases, MeEE2 have been reported (Lai et al., 2000; Yu and Huang, 2005; Bonin and Simpson, 2007). Sorption of EE2 to kaolinite and montmorillonite was reduced 50% in the presence of E2 and E1. Similarly, log K_f values for sorption of EE2 to soil were reduced in the presence of E2 and E1 for EE2 concentrations of 100 - 450 mg/L (Bonin and Simpson, 2007). At low EE2 concentrations, the presence of phenanthrene reduced EE2 sorption in soil by as much as 35% at phenanthrene concentrations of 10 - 100 micrograms per liter (µg/L). This effect decreased as the concentration of EE2 increased (Yu and Huang, 2005). Sorption of EE2 and MeEE2 was reduced by more than 50% and 31%, respectively, in the presence of estradiol valerate at estrogen concentrations of 100 µg/L (Lai et al., 2000). These studies show that the presence of co-solutes can impact the mobility of steroidal chemicals in biosolids-amended soils.

5.3.3 Brominated Flame Retardants

This study focuses on several of the brominated flame retardants (BFRs) known as polybrominated diphenyl ethers (PBDEs) as well as tetrabromobisphenol A (TBBPA). TBBPA accounts for more than 50% of BFRs used (Sun et al., 2008a; Sun et al., 2008b; Sun et al., 2008c) and PBDEs account for approximately 30% of BFR consumption (Palm et al., 2002). Of the 209 PBDE congeners, only a subset have been widely used for commercial applications and the penta-, octa-, and decabrominated PBDEs predominate (Palm et al., 2002; Gouin and Harner, 2003). The BFRs are generally predicted to sorb to soils according to the hydrophobic partitioning paradigm, though the pK_a values of TBBPA (pK_a1 and pK_a2 values of 7.5 and 8.5, respectively; Sun et al., 2008b; Sun et al., 2008c) suggest that pH-dependent sorption could be expected. Log K_{ow} values for PBDEs have been reported by various authors (Palm et al., 2002; Gouin and Harner, 2003; Kuramochi et al., 2007) and are summarized in Table 5-1. Log K_{ow} values increase with increasing bromine content and range from 5.03 (BDE 8/11) to 11.15 (BDE 209). The log K_{ow} values reported for TBBPA ranged from 0.34 - 5.34 (Xia et al., 2005; Sun et al., 2008b; Sun et al., 2008c).

Sorption Data

Few studies of BFR sorption in soil have been conducted. Reported K_d values were found only for penta BDEs and TBBPA. Log K_d values for penta BDEs ranged from 1.95 in sandy soils to 4.65 in humus rich soils (Litz, 2002). The log K_{oc} value for the humus-rich substrate was 5.09. The increase in K_d values from sandy to humus-rich substrates suggests that sorption is strongly related to soil f_{oc}, which is consistent the hydrophobic partitioning paradigm. Other studies by Sun et al. (2008b; 2008c) found log K_d values of 1.27 - 2.6 in sorption experiments with TBBPA in soils ranging from sandy to silty loam. As with the penta BDEs, the log K_d values increase with increasing f_{oc}. To further evaluate the impact of f_{oc} on TBBPA sorption, sorption studies were conducted in soils that had been combusted to remove the majority of organic matter. Sorption of TBBPA was reduced by 90% in combusted soils, suggesting that organic matter plays the main role in sorption for this chemical (Sun et al., 2008c).

Experiments were also completed investigating the impacts of DOM on the sorption of TBBPA to sandy loam, loamy clay, and silt loam (Sun et al., 2008b). DOM may impact solute-soil interactions and can affect sorption by introducing solute-DOM and DOM-soil interactions. For example, DOM may sorb to soil and increase f_{oc} thereby increasing sorption of solutes. Conversely, DOM may complex with solutes and decrease their sorption in soil. The effects of DOM on TBBPA sorption varied from an increase in sorption in a sandy loam soil to a decrease of sorption in a silt loam soil. Addition of DOM increased solution pH, which may have also altered the speciation of TBBPA. To separate the impacts of pH and DOM on the sorption of TBBPA, the authors compared K_d values in the presence and absence of DOM at the same pH. The results suggest that DOM causes an increase in TBBPA sorption, which is likely due to a DOM-induced increase in the f_{oc} of the soil. At higher DOM concentrations, the degree of sorption enhancement decreases. When TBBPA is bound to the DOM in solution this will cause an apparent increase in dissolved TBBPA and potentially reduce the amount of TBBPA available for sorption (Sun et al., 2008b).

Effects of pH and Solution Chemistry on Sorption

Though pH-dependent sorption of PBDEs is not expected, TBBPA is expected to exhibit pH dependent sorption. Sorption experiments at varying pH showed TBBPA sorption decreases above pH 6. At higher pH, the anionic form of TBBPA increases, leading to electrostatic repulsion between TBBPA and negatively charged soil surfaces (Sun et al., 2008c). This study also investigated the impacts of ionic strength on the sorption of TBBPA in loamy clay and silt loam soils. Ionic strength was adjusted by the addition of Ca^{2+} at concentrations of 0.001M, 0.01M, and 0.1M. Sorption increased with increasing ionic strength, with sorption coefficients increasing 28.5 - 315% depending on the soil type and ionic strength. However, ionic strength can impact solution pH, and so TBBPA sorption was investigated at three different ionic strengths while holding pH constant. There was no significant difference in sorption at the three different ionic strengths when the solution pH was held constant, suggesting that pH was the primary cause of the observed sorption differences (Sun et al., 2008c).

5.3.4 Perfluorochemicals and Perfluorochemical Precursors

Perfluorochemicals and PFC precursors included in this study are listed in Table 2-1. These include endproduct PFCs, which typically exist as anionic surfactants at environmentally-relevant pH values, as well as PFC precursors such as fluorotelomer alcohols (FTOHs). No sorption data were available for PFC precursors such as the perfluoroalkyl phosphoric acids (PAPs). Limited sorption data exist for many PFCs and PFC precursors, though when data are not available, some inferences can be made from structurally analogous chemicals. For example, the structural differences between many PFCs and PFC precursors are primarily due to the length of the perfluoroalkyl chain. The addition of a CF_2 group tends to alter the mobility (i.e., increase sorption) of these chemicals in a linear fashion.

PFCs and PFC precursors are expected to sorb via different mechanisms, owing to the fact that many PFCs exist as anions at circumneutral pH values, whereas PFC precursors such as FTOHs are neutral. Thus, one might expect FTOHs to sorb via the hydrophobic organic matter partitioning paradigm. In fact, the neutral PFC precursor 8:2 FTOH demonstrated relatively consistent log K_{oc} values in five soils as well as little correlation between sorption and other soil properties (Liu and Lee, 2005). Not surprisingly, log K_{ow} values of FTOHs increase with increasing chain length. Reported log K_{ow} values ranged from 3.28-3.3, 4.54-4.7, 5.58-6.14, and 7.57 for 4:2, 6:2, 8:2, and 10:2 FTOHs, respectively (Arp et al., 2006; Carmosini and Lee,

2008a). This suggests that sorption may increase with increasing chain length and that trends in transport of these chemicals may depend in part on chain length and resulting hydrophobicity.

PFCs such as perfluorinated carboxylic acids (PFCAs) are generally expected to have low pK_a values, such that only the anionic forms are present at environmentally-relevant pH values (Goss, 2008). This assumption is important since neutral forms of these chemicals would potentially display different transport than the anionic species due to higher volatility and higher sorption potential (Goss, 2008). The pK_a values of various PFCAs were predicted using SPARC and COMO-RS (Goss, 2008). The study does not pinpoint specific pK_a values for these chemicals, but generally reports low pK_a values such as -0.5 for perfluorooctanoate (PFOA) (Goss, 2008). However, as suggested by the authors (Goss, 2008), there appears to be some uncertainty regarding pK_a values for some PFCs. For example, others have reported the pK_a of PFOA measured using a water-methanol mixed solvent approach to be 2.3 for higher concentrations of PFOA and 3.8 at lower concentrations (Burns et al., 2008). Obviously, the pK_a would have a significant impact on the fraction of anionic PFCs present at environmentally relevant pH values. In sorption experiments, amounts of PFC in the neutral form are typically assumed to be negligible. Further pK_a data for PFCs may be needed to determine if this assumption is valid (Burns et al., 2008).

Studies suggest that both hydrophobic and electrostatic effects contribute to the sorption of anionic PFCs. Sorption of anionic PFCs was strongly correlated with sediment or soil f_{oc}, even when significant quantities of iron oxide (positively charged mineral surface) were present (3M, 2000; Liu and Lee, 2005; Higgins and Luthy, 2006; Johnson et al., 2007). This does not mean, however, that electrostatic interactions are absent from the sorption process. Increasing sorption of anionic PFCs to sediment with decreasing pH and increasing Ca^{2+} concentrations (Higgins and Luthy, 2006; Johnson et al., 2007) suggest that electrostatic interactions affect sorption (Higgins and Luthy, 2006). The role of electrostatic interactions could increase in environments low in organic matter (Johnson et al., 2007). Additionally, modeling efforts seems to support that both hydrophobic and electrostatic effects contribute to overall sorption of anionic PFCs (Higgins and Luthy, 2007).

Sorption Data

Few studies are available discussing the sorption of PFCs. Sorption of the PFC precursor 8:2 FTOH was studied in five soils and found to be linear in all cases (Liu and Lee, 2005). Sorption studies were also completed with sediment and the following anionic PFCs and PFC precursors: PFOA, perfluorononanoate (PFNA), perfluorodecanoate (PFDA), perfluoro undecanoate (PFUnDA), perfluoroctanesulfonate (PFOS), perfluorodecanesulfonate (PFDS), 2-(*N*-methylperfluorooctanesulfanamido) acetic acid (N-MeFOSAA), and 2-(*N*-ethyl perfluorooctanesulfonamido) acetic acid (N-EtFOSAA; Higgins and Luthy, 2006). Sorption of these chemicals was mostly linear. For anionic PFCs and PFC precursors, log K_d values for natural sediments ranged from 0.41 for PFOA to 2.47 for N-EtFOSAA (3M, 2000; Higgins and Luthy, 2006; Johnson et al., 2007). Log K_d values for FTOHs range from -0.15 for 4:2 FTOH to 0.79 for 10:2 FTOH (Liu and Lee, 2007).

Studies of both FTOHs (Liu and Lee, 2005) and anionic PFCs (Higgins and Luthy, 2006) show that sorption increases with increasing chain length. Sorption of FTOHs increased 0.87 log units with the addition of each CF_2 moiety (Liu and Lee, 2007), whereas sorption of the anionic PFCs listed increased 0.5 - 0.6 log units with the addition of each CF_2 moiety (Higgins and

Luthy, 2006). In the latter study, the increase in K_d with increasing chain length was consistent among different PFC subclasses and different sediments.

DOM can also impact sorption of FTOHs. Both calculated and measured DOM sorption coefficients for 8:2 FTOH show that the DOM has the potential to reduce sorption, increase S_w, and potentially increase mobility (Liu and Lee, 2005; Carmosini and Lee, 2008a), depending on source of organic carbon and the chain length of the FTOH. Sorption of 8:2 FTOH to various sources of DOM showed that DOM from soil enhanced S_w the most and DOM originating from biosolids enhanced S_w the least (Carmosini and Lee, 2008a). DOM had little impact on shorter chain FTOHs sorption (Carmosini and Lee, 2008a). One study found evidence of an irreversible process during FTOH sorption experiments as evidenced by a reduced extraction efficiency of 8:2 FTOH over a 3-day time period (Liu and Lee, 2005). This suggests limited mobility of FTOHs under many conditions, though mobility may be enhanced in the presence of high concentrations of DOM.

Electrostatic Effects on Sorption

Sorption of anionic PFCs is documented to increase with decreasing pH. Higgins and Luthy (2006) found sorption of various anionic PFCs to sediment increased with decreasing pH from pH 7.5 to 5.7. Similarly, PFOS sorption to goethite and kaolinite decreased with increasing pH (Johnson et al., 2007). These trends support the idea that electrostatic interactions play a role in PFC sorption (Higgins and Luthy, 2006; Johnson et al., 2007). The former study also examined PFC sorption with changing ion concentrations (Higgins and Luthy, 2006). Sorption of anionic PFCs to sediment increased with increasing Ca^{2+} concentration. However, this trend is not thought to be an ionic strength effect due to the fact that, though it resulted in a similar increase in ionic strength, increasing the Na^{+} concentration did not impact the sorption of the chemicals (Higgins and Luthy, 2006).

Effects of Co-Solutes on Sorption

Sorption experiments conducted using a solution with various anionic PFCs and sediment showed somewhat nonlinear isotherms for many of the chemicals measured. Nonlinear sorption can be attributed to competitive sorption of multiple chemicals, but sorption experiments conducted with PFOS alone showed no difference in sorption from experiments completed in the presence of PFCs. Therefore, competitive sorption was not likely to be the cause of the nonlinearity of the isotherms (Higgins and Luthy, 2006).

Sorption Modeling

Four types of software (EPI Suite, ClogP, SPARC, and COSMOtherm) were used to predict various partition coefficients including K_{ow} values for select PFC precursors (Arp et al., 2006). Predicted values were then compared to experimental values to determine the accuracy of the model. SPARC and COSMOtherm performed the best of the four models tested; however no model was able to predict all partitioning constants within the goal of an order of magnitude. Therefore, results stressed the importance of experimental data (Arp et al., 2006). Using experimental data, one study derived a model to predict sorption of anionic PFCs to sediment (Higgins and Luthy, 2007). The model was developed assuming that both hydrophobic and electrostatic effects contributed to sorption. Gibbs free energy terms were used to estimate the contributions of each, with the resulting model fitted the experimental data within a factor of two. However, the model was unable to duplicate the experimental nonlinearity observed.

5.3.5 Plasticizers

The only plasticizer included as a high priority TOrC in this study is bisphenol A (BPA). Few data are available regarding the mobility of BPA in the environment, though BPA is generally expected to sorb via hydrophobic organic matter partitioning. Reported log K_{ow} values range from 3.3 - 3.6 (Ying and Kookana, 2005; Tan et al., 2007). As BPA's pK_a is between 9.59 and 10.2 (Pan et al., 2009), only very high soil pH values are expected to affect sorption.

Sorption Data

Sorption of BPA in soil is essentially linear (Ying et al., 2003) and increases with increasing f_{oc} (Ying and Kookana, 2005). Others have investigated the sorption of BPA onto kaolinite montmorillonite, and goethite (Shareef et al., 2006). Partitioning coefficients were not reported, though the highest sorption was to montmorillonite, and the lowest sorption was to kaolinite (Shareef et al., 2006). This work also included a study of the pH-dependence of BPA sorption over pH range 4 - 10. Only sorption to montmorillonite was pH-dependent and increased with increasing pH. The authors noted that over the pH range studied, the surface charge of all three minerals would have changed. Therefore, since sorption to two of the three minerals was independent of pH, sorption likely does not depend on surface charge (Shareef et al., 2006). However, since this would likely apply to all three minerals, the authors ruled out surface charge as a factor as well. Instead, the authors attribute the pH-dependent sorption of BPA to montmorillonite to the flocculation of the mineral, which changes with pH. At higher pH values, the flocs are arranged such that they would allow increased access of BPA to the clay interlayers (Shareef et al., 2006).

5.3.6 Surfactants

The surfactants 4-cumylphenol and 4-tert-octylphenol were included as part of this study. Little information was found for 4-cumylphenol outside of octanol-water partitioning data. Despite being relatively soluble, 4-tert-octylphenol has reported log K_{ow} values of 4.12 - 5.85 (Ahel and Giger, 1993; Yamamoto et al., 2003; Kawaguchi et al., 2004; Ying and Kookana, 2005; Tan et al., 2007; Tan et al., 2008). These values suggest that 4-tert-octylphenol will have a strong affinity for the solid phase. 4-cumylphenol is reported to have a similar log K_{ow} value of 4.1 (Tan et al., 2007; Tan et al., 2008). As surfactants are known to exhibit unique sorptive behavior, it is difficult to predict the dominant processes affecting the sorption of these TOrCs in biosolids-amended soils, though organic matter likely plays some role in their sorption.

Sorption Data

The majority of studies of sorption of 4-tert-octylphenol have been done using sediment from sources such as rivers and springs (Johnson et al., 1998; Zhou, 2006; Navarro et al., 2009) with few studies involving soil (Ying and Kookana, 2005). Log K_d values for sorption of 4-tert-octylphenol to soil and sediment range from 0.78 - 3.25 (Johnson et al., 1998; Ying and Kookana, 2005; Zhou, 2006). Recent studies of sorption of 4-tert-octylphenol to spring sediments find that sorption was somewhat nonlinear (Navarro et al., 2009). Sorption of 4-tert-octylphenol to soil and sediment increased with increasing f_{oc} (Johnson et al., 1998; Ying and Kookana, 2005). Despite large variations in K_d values for 4-tert-octylphenol, log K_{oc} values are consistent (Zhou, 2006), suggesting that f_{oc} is the dominant factor in controlling sorption of this chemical (Zhou, 2006).

Sorption experiments of 4-tert-octylphenol to DOM revealed no correlation between log K_{dom} and log K_{ow} (Yamamoto et al., 2003), suggesting that hydrophobic interactions are not the primary driver of sorption to DOM. However, the authors recommended verification of the trend with techniques other than the fluorescence quenching utilized to measure the log K_{dom} values (Yamamoto et al., 2003). In addition, one study looked at the sorption of 4-tert-octylphenol to sediment in the presence of nonylphenol (Navarro et al., 2009). Sorption of 4-tert-octylphenol was slightly reduced in the presence of nonylphenol, suggesting these two surfactants may compete for sorption sites.

Desorption of 4-tert-octylphenol from sediment depended on how long the compound had been initially sorbed to the sediment and the length of desorption time (Zhou, 2006). The results were consistent with observed sorption kinetics of fast initial sorption of the compound to the sediment surface, followed by slower diffusion into pore spaces. More desorption in a shorter time frame would be expected from sediments to which 4-tert-octylphenol had only undergone sorption to the sediment surface. Likewise, less desorption at a slower rate would be expected from those sediments to which 4-tert-octylphenol had diffused into pore spaces. These experiments show that this compound can be expected to be more sorptive the longer it is in contact with sediment (Zhou, 2006).

5.4 Volatilization of Biosolids-Borne Trace Organic Chemicals in Soils

Volatilization is not expected to be a significant process leading to mobilization of the majority of target TOrCs from biosolids-amended soils, as TOrCs that are particularly volatile are not expected to accumulate in biosolids in the first place. Moreover, volatility is most important for the neutral TOrCs, which are generally expected to be strongly associated with the biosolids-derived organic matter and/or soil organic matter. However, volatilization may be an important process for the PBDEs, a sub-class of the BFRs, and the FTOHs, a sub-class of PFC precursors.

5.4.1 Polybrominated Biphenyl Ethers

Modeling studies suggest a potential for lower congener PBDEs (BDE 15 and BDE 28) to highly volatile and capable of being transported long distances through the atmosphere (Gouin and Harner, 2003). Similarly, (Palm et al., 2002) calculated partitioning estimates for BDE 47, BDE 99, and BDE 209 that predicted BDE 47 had the greatest potential for atmospheric transport among the congeners, as would be expected.

Atmospheric transport of PBDEs can experience seasonal fluctuations. Harner and Shoeib (2002) reported that octanol-air partition coefficients (K_{oa}) for 13 PBDEs increased with decreasing temperature, suggesting that PBDEs should partition more strongly into solid phases at colder temperatures. Gouin and Harner (2003) noted that this temperature-dependence can lead to a rapid influx of PBDEs into the atmosphere as temperatures increase in the spring and release PBDEs stored in snowmelt during the winter. Multiple deposition and release cycles could result in particularly long range transport of these TOrCs (Gouin and Harner, 2003).

Two studies were found that used partitioning coefficients in a four-compartment model consisting of soil, sediment, air, and water, to predict partitioning of PBDEs (Tittlemier et al., 2002; Gouin and Harner, 2003). Both studies found that PBDEs partition primarily to the soil and sediment phases, and that lighter congener PBDEs partition to air more than the heavy congeners. The results are in agreement a similar study that included suspended sediment and fish (Palm et al., 2002). PBDEs partitioned primarily to soil and sediment phases and lighter congener PBDEs partitioned to the atmosphere. The model results are consistent with the log

K_{ow} trends discussed above, though these simulations do not account for the potential debromination of higher congener PBDEs to lower congeners (Tittlemier et al., 2002).

5.4.2 Fluorotelomer Alcohols

If perfluoroalkyl polymers are present in biosolids and undergo degradation as some evidence suggests (Washington et al., 2009), FTOHs would likely be released into the soil environment. Despite the considerable evidence pointing to global transport of FTOHs via the atmosphere (Shoeib et al., 2006; Stock et al., 2007), experimental or modeling studies examining the release of FTOHs from soil are limited. Biodegradation studies of ^{14}C FTOHs in both soil (Wang et al., 2009) and activated sludge (Wang et al., 2005b) microcosms revealed considerable transfer of ^{14}C volatile organics to the headspace of the microcosms. Transfer was diminished in the live systems, presumably due to rapid biotransformation of the FTOHs to non-volatile species (Wang et al., 2009).

As a measure of potential partitioning to the atmosphere, several studies have measured and modeled both Henry's Law constants (K_{aw} values) and K_{oa} values (Lei et al., 2004; Goss et al., 2006; Thuens et al., 2008). Unfortunately, some of the reported values differ by an order of magnitude (Thuens et al., 2008), though the two most recent studies appear to be in much more agreement with each other than the earlier study. Increasing log K_{oa} values with increasing FTOH perfluoroalkyl chain length have been reported, and range from 4.57 to 6.20 (at 25°C) for 4:2 FTOH to 12:2 FTOH (Thuens et al., 2008). Whether significant volatilization of biosolids-borne FTOHs would occur under field conditions has not yet been evaluated, and may be a data gap that should be addressed if FTOHs appear to be stable and present in biosolids-amended soils.

5.5 Field Observations

The majority of the discussion surrounding the mobility of biosolids-borne TOrCs has focused on measured or predicted sorption coefficients and, for a small subset of TOrCs, coefficients related to their potential volatilization from biosolids-amended soils. The following section addresses data collected from larger-scale studies, whether at the soil-column scale or the field scale. No field or column data were found for cimetidine, PFCs and PFC precursors, or the targeted plasticizers and surfactants. In some cases, important parameters such as soil type, pH, CEC, and moisture and organic matter content were not available from the published studies. Such data are crucial for making broader conclusions as to the mobility of TOrCs in biosolids-amended soils. Given the variety of biosolids application techniques (i.e., broadcast application, injection) and the types of biosolids applied (dewatered vs. liquid), drawing broad conclusions from these field observations is problematic, particularly when important field parameters are not reported.

5.5.1 Pharmaceuticals and Personal Care Products

5.5.1.1 Antimicrobial Agents

Laboratory studies indicate that the antimicrobial agents TCS and TCC undergo sorption to solids and that TCC is more sorptive than TCS. Research has been completed looking at the fate of these TOrCs in column studies (Snyder, 2009), runoff, and tile drainage following land application of biosolids (Lapen et al., 2008b; Topp et al., 2008a; Topp et al., 2008b; Edwards et al., 2009; Sabourin et al., 2009).

Column Studies

Column studies were completed to determine the leachability of TCC in biosolids-amended soils (Snyder, 2009). Soils amended with one of 11 biosolids at a rate equivalent to 18 - 52 Mg/ha were packed into 17 cm by 5 cm columns. Columns were leached bi-weekly for 14 weeks with a final leaching at 5.5 months. Approximately four pore volumes of leachate were collected by the end of the study. TCC was detected in leachate from treatments involving 8 of the 11 biosolids studied, but concentrations were low. In biosolids for which the initial concentration of TCC was known, 0.02 - 0.18% of the TCC applied was collected in leachate (Snyder, 2009).

Tile Drainage

Two studies were found investigating the impacts of biosolids application on tile drainage in agricultural fields (Lapen et al., 2008b; Edwards et al., 2009). One study focused on PPCP concentrations following application of liquid municipal biosolids (Lapen et al., 2008b), and the second focused on concentrations following application of dewatered municipal biosolids (Edwards et al., 2009). Both studies utilized 100 m by 15 m plots centered over tile lines and the plots subjected to natural meteorological conditions. Tile drains were 0.8 m below the surface. Following application of liquid municipal biosolids at a rate of 93,500 L/ha, PPCP concentrations, including TCS, were detected within minutes of application. Liquid municipal biosolids had been spiked with TCS at 3,872 ng/L. The initial detection of TCS is thought to be the result of flow through macropores such as worm boreholes. TCS continued to be detected in tile drainage for several weeks following application (Lapen et al., 2008b). A subsequent study in the same study area showed that residual concentrations of TCS were present in the tile drainage nine months following application of liquid municipal biosolids (Edwards et al., 2009). The subsequent study focused on the application of dewatered municipal biosolids. As mentioned, PPCP concentrations were measured to evaluate any residual concentrations from application of the liquid biosolids. Dewatered municipal biosolids containing 14,000 ng/g_{dw} and 8,000 ng/g_{dw} of TCS and TCC, respectively, were applied at a rate of 8 Mt dw/ha. Average TCS concentrations in tile water following application of dewatered biosolids were higher than average TCC (43 ng/L vs. 0.73 ng/L). This trend agrees with sorption data of these chemicals, which show stronger sorption of TCC.

Runoff

Similar to tile drainage, two studies were completed examining TCS and TCC in runoff following application of liquid municipal biosolids (Topp et al., 2008b) and dewatered municipal biosolids (Sabourin et al., 2009). Liquid municipal biosolids were applied to a 15 m by 40 m plot via two methods (broadcast application and injection) and subjected to simulated rainfall until 10 L of runoff was collected. Chemical concentrations were greater in runoff from fields receiving broadcast application of biosolids than in those receiving biosolids injections. Concentrations of TCS were detected in runoff up to nine months following application of the biosolids. The concentration of TCS in the runoff was 258 ng/L the day following broadcast application (Topp et al., 2008b). Similar studies were completed using dewatered municipal biosolids applied to 2 m^2 microplots that received simulated rainfall at days 1, 3, 7, 21, 34 until 10 L of runoff was collected (Sabourin et al., 2009). The maximum detected concentration of TCS (109.7 ng/L) occurred one day after application. In general, 40 times more TCS was exported via runoff than TCC (Sabourin et al., 2009). This is consistent with both sorption data and studies of TCS and TCC concentrations in tile drainage.

5.5.1.2 Tetracycline Antibiotics

One study used soil columns to measure the leaching potential of OTC (Rabolle and Spliid, 2000). Due to the similarities between TC and OTC, it is expected that both TOrCs would show similar results. The column studies were conducted in 30 cm of soil, 5.2 cm in diameter packed with sandy loam or sand soil. A 1 mL solution of 500 mg/L OTC was applied followed by the equivalent of 500 mm of simulated rainfall. OTC was not detected in leachate from these columns suggesting that the leaching potential of tetracyclines is low (Rabolle and Spliid, 2000).

5.5.1.3 Fluoroquinolone Antibiotics

One study was found that used lysimeters to measure the leaching potential of ofloxacin (Drillia et al., 2005). Due to the similarities between ofloxacin and CIP, it is expected that both TOrCs would show similar results. The column studies were conducted in columns 20 cm by 4.4 cm. Lysimeters were packed with 10 cm of one of two soil types. One soil type was high in f_{oc} and low in clay, the other low in f_{oc} and high in clay. Lysimeters received either 0.5 or 1 L of a solution containing 20 mg/L ofloxacin followed by varying rates of simulated rainfall, depending on soil type. Ofloxacin was not detected in leachate from either soil type (Drillia et al., 2005). Another study measured concentration versus depth profiles of CIP in an experimental field site after application of biosolids (Golet et al., 2003). Biosolids were applied to an agricultural field at a rate of 50 t/ha, ten times the amount allowed in Switzerland, to simulate a worst-case scenario. CIP movement was limited to the top 2.5 cm of soil during the first five months of the study. In months 5 through 21, concentrations from 2.5 -12 cm increased. However, these concentrations were extremely low, falling between the limit of quantitation and the limit of detection. Studies of both the depth profiles and lysimeter leachate indicate that transport potential of fluoroquinolone antibiotics is low.

5.5.1.4 Synthetic Musks

Sewage sludge spiked with 10 mg/kg HHCB and AHTN was applied to lysimeters 14 cm in diameter by 30 cm in height and packed with Luvisol, Cambisol, or Podzol soils. Lysimeters were treated with 200 mm of simulated rainfall to study leaching potential of these two chemicals. HHCB and AHTN were detected only in the leachate from columns packed with the Podzol. The concentration of HHCB decreased from an initial leachate concentration of 0.28 to 0.19 μg/L. Similarly, the concentration of AHTN decreased from 0.26 to 0.06 μg/L. Leachate concentrations of HHCB and AHTN were thought to be the result of preferential flow (Litz et al., 2007). Following completion of the column studies, the soil was divided into six intervals and analyzed for HHCB and AHTN. Results showed that >80% of the HHCB and >75% of the AHTN was retained in the sewage sludge layer on the top of the soil columns (Litz et al., 2007).

5.5.6 Steroidal Chemicals

The leaching potential of hormones in soil following land application of pig slurry via direct injection was examined (Laegdsmand et al., 2009). The objective of the study was to look at leaching of 17-α-estradiol, E2, and E1; however, EE2 was used as a reactive tracer. Leaching of the hormones was studied in lysimeters 60 cm in diameter and 100 cm long and packed with undisturbed soil cores of either loamy or sandy soil. The slurry was spiked with 92.2 mg/kg wet weight of EE2 and applied in a manner that simulated direct injection at a rate of 30 tons/hectare (Laegdsmand et al., 2009). Two experiments were completed with the lysimeters. The first was an irrigation experiment that applied artificial rainfall at a rate of 10mm/hr for 12 hrs. The second was a field experiment that exposed the lysimeters to outdoor, natural rainfall conditions for 16 weeks. Following completion of the irrigation experiment, 0.0015-0.0027% and 0.0002 - 0.0005% of the total EE2 applied to the soil had leached from the lysimeters containing loamy soil and sandy soil, respectively. Following the end of the field experiment at 16 weeks, a total of 0.002 - 0.0031% and 0.0003 - 0.0011% of EE2 had leached from the lysimeters containing loamy and sandy soil, respectively. During both experiments more EE2 leached from the loamy soil (Laegdsmand et al., 2009), but leaching losses of both compounds was minimal.

5.5.7 Brominated Flame Retardants

Leaching of penta BDEs from soil lysimeters 10 cm in diameter and 25 cm in height packed with either Histosol or Cambisol was examined (Litz, 2002). Soils were spiked with 455 μg/kg of penta BDEs and water was applied to the surface at a rate of 285 mL/hr for the Histosol and 441 mL/hr for the Cambisol. Minimal leaching was observed, but some breakthrough of PBDEs occurred in both substrates early in the experiment and was attributed to preferential flow that occurred prior to establishing sorption equilibrium (Litz, 2002). Penta PBDEs consist of heavier (more brominated) PBDEs (i.e. BDE 85, BDE 100, and BDE 105). As previously discussed, log K_{ow} values increase with increased bromination; therefore, leaching of lower brominated PBDE congeners with lower log K_{ow} values may be of concern. These data (Litz, 2002) suggest that transfer of higher brominated PBDE congeners to groundwater is not likely.

5.6 Conclusions

The objective of this chapter was to understand the potential for TOrC mobility in the environment, especially when introduced through biosolids-amended soils. Because of the limited amount of information specific to biosolids-amendment, studies of the mobility of TOrCs in other solid matrices such as soil and sediment were also included. To facilitate an evaluation of TOrC mobility the following were considered: physicochemical data, field studies, and in limited cases, volatilization. In particular, an evaluation of physicochemical data included pK_a values and sorption parameters (i.e. K_{ow}, K_d, K_{oc}). This information has been used to identify where further research is needed to fully understand and predict the mobility of TOrCs in the environment. In all cases, there was a great deal of variability in the data.

With the noted exceptions of the tetracycline and fluoroquinolone antibiotics, the primary sorption mechanism for most of the targeted TOrCs is hydrophobic partitioning to organic matter. As a result, log K_{ow} values are important parameters for predicting sorption of moderate to high K_{ow} TOrCs such as the PBDEs or synthetic musks. In other cases, pH-dependent organic matter partitioning was evident or suspected (i.e., TBBPA, TCS), and this was particularly important for the TOrCs that will only exists as charged species at environmentally-relevant pH values (i.e., PFCs). For the tetracycline and fluoroquinolone antibiotics, other mechanisms such as cation exchange, surface complexation, and cation bridging can provide fairly strong sorption of the TOrCs observed in soils. As these TOrCs have low log K_{ow} values, the use of inappropriate models (i.e., hydrophobic partitioning models) would significantly overestimate the mobility of these TOrCs in soils. Several models have been developed for the tetracycline and fluoroquinolone antibiotics that should enable mobility predictions if sufficient data are available with respect to the geochemical conditions present in biosolids-amended soils. However, very few studies have examined the effects on sorption of the complex geochemical interactions that occur in soil as a result of biosolids-amendment.

To fully understand the mobility of a chemical in the environment it is important to also look at work that has been completed at levels above laboratory-scale sorption experiments such as bench scale column studies and field experiments. Such studies were identified for three compound classes: PPCPs, steroidal chemicals, and BFRs. Studies of TCC and TCS in columns, tile drainage, and runoff all identified the compounds present in leachate. Column studies of tetracycline antibiotics, fluoroquinolones, and synthetic musks found low potential for leaching. Similarly, leaching potential for EE2 was also found to be low despite that the study was conducted with pig slurry, which would be expected to have higher EE2 concentrations than municipal biosolids. Finally, column studies investigating the leaching potential of penta BDEs found leaching potential of these chemicals to be low as well.

5.6.1 Data Gaps

Though significant work has been done to understand the mobility and sorption of TOrCs in the environment, several areas of research still need to be addressed. Sorption coefficients were identified for many TOrCs, but data were not complete for any compound class or subclass. In particular, sorption parameters were missing for many of the BFRs and PFCs. Furthermore, data were extremely limited in all compound classes with respect to larger scale studies at either the bench or field scale. More studies of this nature are needed to fully identify the mobility of biosolids-borne TOrCs, though many studies conclude that leaching potential of compounds is low.

Across the board, more work research is needed that is specific to biosolids-amended soils. Soils that have been amended with biosolids will have different characteristics than those that have not been amended. As shown in the above discussion, characteristics such as pH and f_{oc}, have the potential to impact the mobility of a chemical. Another general factor that needs to be considered is the consistency of the types of studies that have been done and the methods used. If sorption of chemicals is studied in various soil types, it is important that key geochemical parameters be measured (i.e. f_{oc}, CEC, pH) to enable meaningful interpretations of the resultant data. Moreover, at the field-scale, it is important that details regarding the type of biosolids applied, the method of application, the type of agricultural field, and the general hydrologic conditions (i.e., moisture content at time of application) are all collected and included in the published reports. A fourth general issue that extends to all compound classes is the issue of desorption. The limited desorption information available suggests that sorption is not always fully reversible, indicating that models that assume complete reversibility may not always be valid. Further studies of desorption in all compound classes are required to identify which chemicals this issue impacts.

Table 5-2 provides a summary of the mobility data availability with respect to the high priority TOrCs included in this study. The decision used to bin the TOrC classes or subclasses are provided, though in some cases, expert judgment was required to consider data that did not readily fit within the framework developed. As is clear from the table, while a substantial body of knowledge exists regarding the mobility of many TOrCs in soils, some significant data gaps are still evident for some compound classes.

Table 5-2. Summary of Mobility Data Availability for the High Priority TOrCs.

Chemical Class	Data Availability
BFRs	Tier 1
PFCs and PFC Precursors	Tier 2
PPCPs: Antimicrobials	Tier 2
PPCPs: Antibiotics	Tier 2
PPCPs: Synthetic Musks	Tier 2
PPCPs: Other	Tier 0
Plasticizers	Tier 2
Steroidal Chemicals	Tier 2
Surfactants	Tier 2

Data Availability Ranking Decision Criteria:

Tier 0 (No Data)	Essentially no data were available of this type for this class or subclass of TOrCs, including data that could be used for modeling.
Tier 1	For the majority of TOrCs in this class or subclass, physicochemical parameters have been measured (i.e., K_{OW}) that would enable sorption/mobility predications to be made using appropriate models.
Tier 2	For the majority of TOrCs in this class or subclass, mobility has been evaluated in laboratory-based spiking studies employing actual soils or sediments and appropriate analytical protocols.
Tier 3	For the majority of TOrCs in this class or subclass, realistic and nationally-relevant field-scale studies been conducted evaluating the mobility (i.e., leaching) from biosolids-amended soils.

CHAPTER 6.0

BIODEGRADATION OF BIOSOLIDS-BORNE TRACE ORGANIC CHEMICALS IN SOILS

6.1 Introduction

Studies of physical, chemical, and biological processes relating to persistence provide an indication of the current knowledge and data gaps about the ability of biosolids-borne trace organic chemicals (TOrCs) to remain in the environment over time. Possible impacts on attenuation and mobility relating to sorption and potential transport have been addressed in Chapter 5.0. Physical and chemical attenuation processes such as volatilization and photolysis (Vasconcelos et al., 2009) can play a role in attenuation, particularly during surface deployment of biosolids. However, low volatility is a characteristic of the identified compounds of concern and minimal light penetration is expected in mixed soils. For these reasons, this chapter focuses on biosolids-borne TOrC biodegradation and wherever possible, biodegradation in relevant systems.

Understanding the mechanisms by which TOrCs are removed from an environment is necessary for determining persistence and severity of these chemicals, the possible release of intermediate or terminal degradation products, and understanding their short- and long-term environmental impacts. Dissipation includes such physiochemical mechanisms as sorption and partitioning. Biotransformation can result in the production of intermediate or terminal degradation products, which may or may not exhibit toxic effects. Complete oxidation of an organic compound to CO_2 is referred to as mineralization. While the focus of this chapter is on the biodegradation of TOrCs, the available data frequently report a combination of biodegradation and dissipation data and so both are reported here.

The kinetics involved in biodegradation, whether in the laboratory or the natural environment, vary depending on the individual chemicals and the affinity of the degrading organism or community for the contaminant, as well as various environmental factors such as temperature, chemical concentration, and soil-biosolids concentration. Degradation of TOrCs in biosolids-amended soils is believed to be more complex than in un-amended soils due to the unique properties of biosolids. Importantly, TOrCs may tend to be primarily associated with the organic and nutrient-rich biosolids matrix. This imparts complications associated with phase transfer and accessibility; bioavailability and subsequent biodegradation could be impacted with the possibility of enhanced persistence.

Kinetic data, while limited in availability and scope, enable useful predictions of recalcitrance and possible avenues to increase attenuation. In many cases, the available literature relating to biodegradation of the identified compounds was limited to aquatic systems or un-amended soil systems. Admittedly, the extrapolation from aquatic media to soil systems or even from un-amended soils to amended soils is poorly understood. However, these data provide a general picture of relative recalcitrance as well as mechanistic explanations of degradation, possible pathways and intermediate degradation products of concern. Extrapolation of the data to soils, especially biosolids-amended soils, however, is tenuous.

The availability and scope of kinetic data are varied, including such parameters as removal efficiency and half-life measurements. More classical enzymatic models that equate rate of degradation to aqueous concentration have identified half saturation (K_s) and maximal transformation rates (k_{max} or v_{max}) for kinetic fits to the Michaelis-Menten equation. The constants can then be used to determine substrate concentration- (C) dependent rates of transformation (k_c or v_c). This model can be broken into two approximate regimes; lower concentrations regimes can be approximated through pseudo-first-order kinetics (kC), whereas saturation kinetics applicable in higher concentration regimes can be characterized by zero-order kinetics (k). In biosolids and for the TOrCs included in this study (Table 2-1), a simplification can be made by assuming pseudo-first-order, concentration-dependent degradation kinetics due to factors such as phase partitioning and low concentrations. There is an inherent degree of uncertainty with such assumptions, but they provide a baseline for evaluating potential recalcitrance and for identifying data gaps. Most kinetic studies available in the literature assume first-order kinetics.

This chapter reviews data relating to biodegradation of pharmaceuticals and personal care products (PPCPs), steroidal chemicals, brominated flame retardants (BFRs), perfluorochemicals (PFCs) and PFC precursors, plasticizers, and surfactants. As summarized in the conclusions section, some compound classes required using liquid batch studies of pure or mixed microbial cultures rather than data derived from biosolids and/or soils to gain a general perspective on recalcitrance, rates, and degradation products. Wherever possible, more environmentally and topically relevant systems such as aerobic and anaerobic soils, and (most importantly) biosolids-amended soils have been identified and reviewed.

6.2 Pharmaceuticals and Personal Care Products

As a result of widespread manufacture and consumption, numerous antimicrobials are detected in biosolids (Table 4-1). Major environmental contaminants include the widely-used triclosan (TCS), triclocarban (TCC), and the various tetracyclines and fluoroquinolones. Much of the available biodegradation data pertain to these compounds.

6.2.1 Antimicrobial Agents

In a survey of biosolids from 16 wastewater treatment plants (WWTPs), TCC and TCS concentrations ranged from 0.23 to 80 mg/kg and 0.33 to 61 mg/kg, respectively (Xia et al., 2010). In a subsequent analysis of long-term biosolids application on contaminant profile in soils, field plots were treated for 33 consecutive years with 16.8, 33.6, and 67.2 mg biosolids/ha of anaerobically digested primary and waste-activated biosolids, resulting in total mass loads of 554.5, 1109, and 2218 Mg biosolids/ha (Xia et al., 2010). Soil cores sampled from each plot at multiple depths revealed decreasing contaminant concentration with depth. Although initial TCS and TCC concentrations were not defined, the difference between the estimated concentrations in biosolids and the concentrations found in soil after long-term biosolids application suggests that removal was due in part to biodegradation. However, an analysis of soil extractable residues as an indicator of contributory abiotic processes was not performed nor was the possibility explored that compound extractability changed over so many years.

Laboratory studies have been conducted to better understand the biodegradation of TCC in biosolids-amended soil (Snyder, 2009). The mineralization of ^{14}C-TCC to $^{14}CO_2$ was monitored in two types of biosolids-amended soils (silty clay and fine sand). In this study, only slight mineralization (< 5%) was observed and no metabolites were detected, suggesting TCC half-lives are on the order of decades (Snyder, 2009). In another study, soil microcosms containing 100 g of sandy loam were amended with ^{14}C-labeled and unlabeled TCS and TCC via direct application, liquid municipal biosolids or dewatered municipal biosolids (Al-Rajab et al., 2009). Initial TCC and TCS concentrations in the amended soils were 0.33 kBq/g (liquid and dewatered municipal biosolids) and 16.67 Bq/g (direct application) of the labeled compound, and 1 µg/g unlabeled compound (all soils). Under laboratory conditions (30ºC), mineralization of TCS and TCC within the microcosms exhibited similar results: compared to soil amended via direct application, the rate of mineralization of both TCS and TCC in soil + liquid municipal biosolids was faster (Table 6-1). Conversely, mineralization rates decreased with the addition of dewatered municipal biosolids. Al-Rajab and colleagues (2009) also incubated biosolids-amended soil cores in the field at ambient temperatures (10-20ºC). Mineralization and corresponding removal of extractable residues from the soil followed first order kinetics (Table 6-1). Extractable residues underwent an initially rapid decrease in liquid biosolids-amended soils, whereas dewatered biosolids-amended soils experienced a much slower rate of removal compared to direct application.

Table 6-1. First-Order Rate Constants for Mineralization and Removal of Extractable Residues.

Compound	Treatment	r^2	K (d^{-1})	Std error
TCS – Lab Mineralization	Soil	0.9273	0.1666a	0.0191
	Soil + liquid biosolids	0.9565	0.4045	0.0352
	Soil + dewatered biosolids	0.9939	0.0553	0.0018
TCS – Field Extractable Residues	Soil	0.9012	-0.6831	0.1131
	Soil + liquid biosolids	0.9501	-1.1080	0.1795
	Soil + dewatered biosolids	0.4361	-0.2639	0.1501
TCC – Lab Mineralization	Soil	0.9931	-1.0270	0.0351
	Soil + liquid biosolids	0.9929	-2.1270	0.0737
	Soil + dewatered biosolids	0.9429	-0.7299	0.0734
TCC – Field Extractable Residues	Soil	0.0078	0.0231	0.1503
	Soil + liquid biosolids	0.9481	-0.5555	0.0750
	Soil + dewatered biosolids	0.1918	-0.2649	0.3139

Soils were treated directly with TCS and TCC, or with TCC or TCS and via liquid or dewatered biosolids. Data from (Al-Rajab et al., 2009).

In a study comparing TCC and TCS biodegradation in unamended soils (no biosolids were added) under both aerobic and anaerobic conditions, TCS exhibited a half-life of 18 days under aerobic conditions (Ying et al., 2007). Conversely, very little degradation occurred in anaerobic soils (experiments performed in a nitrogen atmosphere) over a period of 70 days; TCS concentration decreased from 1.07 mg/kg to ~0.9 mg/kg suggesting a half-life that may exceed 70 days in unamended soils. In this same study, TCC exhibited an aerobic half-life of 108 days as evidenced by a reduction of an initial TCC concentration of 1.07 mg/kg to 0.63 mg/kg after 70 days. Much like for TCS, TCC degradation was slower under anaerobic conditions; over 70 days an initial concentration of 1.07 mg/kg was reduced to ~0.85 mg/kg.

Limited kinetic data are available for TOrC biodegradation in biosolids-amended soils, and some use activated sludge in bench scale experiments and WWTPs to provide potentially useful reference data for the biodegradation occurring in biosolids-amended soils. The conditions, however, are very different in WWTPs and amended soils, so extrapolation is problematic. In batch experiments with activated sludge, TCS concentration and microbial adaptation were determining factors in TCS mineralization (Federle et al., 2002).

6.2.2 Antibiotics

The sorption and degradation of several antibiotics were tested in aerobically-digested sludge (Wu et al., 2009). Tetracycline (TC) and doxycycline (DTC) were biodegraded while ciprofloxacin (CIP) was not. However, all compounds sorbed strongly and were capable of surviving storage in biosolids and re-entering the environment through land application. In WWTPs in South China, the high-priority ofloxacin and lower-priority norfloxacin were found in high concentrations in secondary sludge, suggesting that these compounds were removed due to sorption and not biodegradation (Xu et al., 2007).

6.2.3 Synthetic Musks

The dissipation of fragrance materials including acetyl-hexamethyl--tetrahydronaphthalene (AHTN) and HHCB were evaluated in biosolids-amended soils (Difrancesco et al., 2004). The study was conducted for one year in four different soils (sandy, silty, clayey, and oxide-rich) amended with biosolids from two WWTPs. Biosolids were applied and thoroughly mixed into the soil. Over the course of the year, concentrations of both chemicals decreased. Concentrations decreased in month 1 through month 3, remained constant in months 3 through 6, and decreased in months 6 through 12. HHCB was not detected by the end of the 12-month period, but AHTN was still present. It is possible for leaching, volatilization, abiotic reactions, and biological transformation to contribute to dissipation of these chemicals. In general, dissipation of AHTN and HHCB was slower in soils with higher organic content. Due to the longevity of AHTN during the study, accumulation of AHTN during repeated biosolids application might be possible, especially in soils with higher organic matter content.

6.3 Steroidal Chemicals

Human hormones are an environmental concern for their widespread occurrence and potential effects on the endocrine systems of aquatic and terrestrial organisms. The chemicals also represent possible intermediates of more recalcitrant synthetic hormone biodegradation. As noted for several other classes of TOrCs, data pertaining to steroidal chemical biodegradation in biosolids-amended soils are scarce. Thus, dissipation and degradation data for a variety of media are reviewed here for potential extrapolation to biosolids systems.

In un-amended soil microcosms, soil moisture content and temperature affected the dissipation and/or mineralization of ^{14}C-17α-ethinyl estradiol (EE2) and ^{14}C-17β-estradiol (E2). At low soil moisture content dissipation of EE2 was slow and did not follow first-order kinetic models (Colucci and Topp, 2001). As moisture content increased, dissipation rates also increased and followed first-order kinetics (Table 6-2). Temperature enhanced EE2 dissipation as evidenced by the decrease in half-life from 7.7 days at 4°C to 3 days at 30°C. An increase in soil moisture content up to field capacity, which ranged from 24-40%, generally increased the rate of E2 removal (Colucci et al., 2001). Temperature was not a significant factor for dissipation but was important for mineralization, with maximum mineralization occurring at 30°C and 37°C. Removal of E2 was accompanied by an increase in ^{14}C-estrone (E1), and the subsequent formation of non-extractable residues was believed to be the result of ^{14}C-estrone biodegradation. First-order dissipation rate constants for E2 and E1 are listed in Table 6-3.

Table 6-2. Dissipation of 10 mg/kg EE2 in Three Soils.

Soil	Moisture (%)	K_D (d^{-1})
Loam	5	0.10
	12	0.14
	20	0.33
	30	0.37
	40	0.22
Silt loam	5	0.03
	15	0.17
	55	0.11
Sandy loam	5	0.08
	10	0.28
	24	0.25

Data from (Colucci and Topp, 2001). Soils adjusted to a range of moisture contents and incubated at 30°C

Table 6-3. Removal of E2 and E1 in Three Agricultural Soils.

Soil	K_D (d^{-1})	
	17β-estradiol (E2)	Estrone (E1)
Loam	2.37	0.75
Sandy loam	3.12	0.41
Silt loam	1.45	1.13

Soils were adjusted to a moisture content of 13% and incubated at 30°C. Soils were supplemented with 1 mg/kg substrate. Data from (Colucci et al., 2001)

In un-amended freshwater sediment cultures, E2 was degraded under methanogenic, sulfate-, iron-, and nitrate-reducing conditions, while EE2 was not (Czajka and Londry, 2006). The rate of E2 dissipation decreased in the following order: iron > sulfate > CO_2 > nitrate. In all environments, E2 was transformed into E1, estriol (E3) and other unidentified, less estrogenic metabolites. Biodegradation of E1, E2, and E3 by *Novosphingobium* sp. ARI-1 also produced no estrogenic metabolites; however, this microorganism was unable to degrade EE2 (Fujii et al., 2002).

The presence of steroidal chemicals such as E2 and EE2 in biosolids suggests that the biodegradation of these compounds in WWTPs and surrounding areas might provide useful data for determining persistence and biodegradation in amended soil environments (Teske and Arnold, 2008). At sewage treatment plants, free estrogens and sulfated estrogens were the dominant species (D'Ascenzo et al., 2003). Removal efficacy of most estrogenic compounds from the conventional activated sludge in South Queensland, Australia wastewater treatment plants ranged from 80-99% (Tan et al., 2007). Upstream of WWTPs, E2, E1, and testosterone were mineralized to CO_2 in stream sediments (Bradley et al., 2009). Downstream, however, testosterone and E1 mineralization was reduced and E2 mineralization doubled compared to levels in upstream sediments. Activated sludge containing nitrifying bacteria was able to degrade EE2 at a rate of 1 μg/g/h when ammonia was the only available energy source while sludge possessing diminished nitrifying capacity was not able to degrade EE2 (Vader et al., 2000). The fates of ^{14}C-labeled estrogen and testosterone were also examined in biosolids obtained from WWTPs (Layton et al., 2000). 84% of ^{14}C-E2, 85% of ^{14}C-E1, and 68% of ^{14}C -testosterone was mineralized to ^{14}C-CO_2. A doubling of temperature had a more pronounced effect on the kinetics of ^{14}C-E2 mineralization than for ^{14}C-testosterone or ^{14}C-EE2 mineralization (Table 6-4).

Table 6-4. Effect of Temperature on Mineralization of Hormones to CO_2.

	First-order rate constant *k* for mineralization to $^{14}CO_2$ (min^{-1})		
Temp	Testosterone	17β-estradiol (E2)	17α-ethinyl estradiol (EE2)
5 - 10°C	0.0161 ± 0.0016 (r^2 = 0.94)	0.0029 ± 0.0002 (r^2 = 0.89)	0.0001 ± 0.0000 (r^2 = 0.98)
22 - 25°C	0.0152 ± 0.0021 (r^2 = 0.77)	0.0042 ± 0.0002 (r^2 = 0.92)	0.0002 ± 0.0000 (r^2 = 0.96)

Data from (Layton et al., 2000)

Amending agricultural soils with slurry swine manure or municipal biosolids has been shown to enhance the biotransformation of E2 to E1 (Jacobsen et al., 2005) while this process occurs slowly in un-amended soils. In other laboratory experiments, four *Rhodococcus* strains (three *R. equi* and one *R. zopfii*) were isolated from activated sludge that were capable of degrading 1 mg of E2, E1, E3, EE2 within 24 h (Yoshimoto et al., 2004). This genus represents common soil bacteria that have been demonstrated to be involved in the biodegradation of a variety of xenobiotic compounds. The lone strain of *R. zopfii* consumed E2 even when glucose was available as a competing carbon source. All four strains decreased the estrogenic activity of E2 to 1% of its normal levels within 24 hours.

6.4 Brominated Flame Retardants

Brominated flame retardants are heavily used in manufacturing, and their widespread presence allows for multiple avenues of entry into the environment. Although the objective of this chapter is to review biodegradation in soil and/or biosolids-amended soils, the majority of BFR research has been conducted using microbiological assays and to a lesser extent WWTP sludge; therefore, it is findings from these studies that are summarized in the remainder of this section to provide relevant extrapolation. Clearly a need exists for further studies pertaining specifically to soil and biosolids-amended soil systems.

Agricultural soil plots which had received subsurface injection of liquid slurry biosolids contained multiple PBDEs including BDE 47, 99, 100, 153, 154, and 183 (Arnold et al., 2008). BDE 47 and 99 were the dominant congeners in the first foot below the surface, ranging from 20 - 60 and 20 - 80 ng BDE/g soil in plots that received 50 and 150 lbs nitrogen per year. The duration and frequency of biosolids application in these studies were not reported, which would be useful for assessing rates of BDE accumulation in soil.

In a study of biosolids from 16 WWTPs, the sum of five PBDE congeners (BDE 47, 99, 100, 153, and 154) ranged from 0.071 to 1.02 mg/kg_{dw} (Xia et al., 2010). In field plots that received a total biosolids application of 2218 mg/ha, PBDE concentration decreased as soil depth increased, from 658 µg/kg (0-15 cm), to 105 µg/kg (15-30 cm), to 4.2 µg/kg (60-120 cm). The initial concentration of PBDEs applied to each field plot, which are not reported, are necessary for determining dissipation or degradation trends in this case.

Many BFRs are brominated diphenyl ethers (BDE) possessing varying numbers of Br atoms. Microbiological assays have shown that chemical speciation and the debrominating culture directly impact BDE biodegradation. For example, in one study *Sulfurospirillum multivorans* debrominated BDE 209 to hepta- and octa-BDEs but was unable to debrominate the octa-BDE mixture. Conversely, *Dehalococcoides*-containing cultures debrominated an octa-BDE mixture but were unable to debrominate BDE 209 (He et al., 2006). It was also found that enriched *Dehalococcoides* cultures produced a number of toxic debromination products such as BDE 154 and BDE 99 (He et al., 2006). In other work, BDEs 47, 99, 153, 183, 196, 197, and 203 were all debrominated to some extent by pure or mixed cultures containing *Dehalococcoides* species, *Dehalobacter restrictus* PER-K23, and *Desulfitobacterium hafniense* PCP-1. All exhibited similar debromination pathways with preferential removal of para and meta bromines (Robrock et al., 2008). Debromination of highly brominated congeners was slower than that of lesser-brominated congeners.

Microbial cultures capable of debrominating tetrabromobisphenol-A (TBBPA) to bisphenol-A (BPA) were less efficient in the presence of the intermediate metabolites monoBBPA, diBBPA, and triBBPA (Arbeli and Ronen, 2003). However, a mixed microbial community isolated from soil spiked with commercially-available PBDEs was able to degrade PBDEs as a sole carbon source (Vonderheide et al., 2006). Denatured gradient gel electrophoresis (DGGE) and Deoxyribonulcelic acid (DNA) sequencing identified the presence of Bacteroidetes, Formicates, Proteobacteria, and Actinobacteria in this PBDE-degrading community.

In digested sewage sludge, TBBPA, hexabromocyclododecane (HBCD), and BDE 209 were all degraded (Gerecke et al., 2006). TBBPA and HBCD had very short half-lives, but the half-life for BDE 209 was close to two years (Table 6-5). The half life for (±)-α-HBCD was nearly twice that of (±)-β-HBCD and (±)-γ-HBCD. Since γ-HBCD is the primary component of technical HBCD mixtures, identifying organisms capable of its degradation will be valuable for understanding HBCD degradation as a whole. From soil contaminated with γ-HBCD (Yamada et al., 2009), 13 strains of bacteria were able to degrade the compound; one Pseudomonad strain (HB01) proved particularly efficient.

Table 6-5. Biodegradation Kinetics of Three BFRs in Digested Sewage Sludge.

BFR	First-order rate constant, k (d^{-1})	Half-life (d)
TBBPA	1.2 ± 0.06	0.59
HBCD	1.1 ± 0.	0.66
BDE 209	1×10^{-3}	7×10^{2}

Data from (Gerecke et al., 2006)

6.5 Perfluorochemicals and PFC Precursors

PFCs comprise a variety of persistent environmental contaminants either used directly in industry or formed as byproducts of other chemicals. Byproducts of fluorotelomer alcohol (FTOH) biodegradation include potentially hazardous compounds such as poly- and perfluorinated acids. Therefore, understanding the degradation of FTOHs is needed to better understand the behavior and toxicity of various byproducts in the environment. In WWTP effluent, concentrations of perfluorooctanoate (PFOA) were higher than in influent, suggesting that biodegradation of precursors, such as FTOHs, contributes to the increase in PFOA concentrations during wastewater treatment processes (Loganathan et al., 2007).

In mixed bacterial cultures enriched from sediment and groundwater, 85% of spiked 8:2 FTOH was degraded within 7 days, and less than 2 μg/L remained after 16 days (Dinglasan et al., 2004). An initial half-life was estimated at ~0.2 day/mg biomass protein, followed by a second half-life of 0.8 day/mg, suggesting complex concentration-dependent kinetics. Three biotransformation products have been identified using ^{14}C-labeling and quadrupole time-of-flight mass spectrometry that collectively represent one-third of the initial ^{14}C mass after 28 days. The masses of these FTOH transformation products were found to be 27%, 6%, and 2% of initial ^{14}C mass, respectively, for 8-2 saturated acid, 8-2 unsaturated acid, and PFOA. 57% of the initial ^{14}C mass remained as the parent compound after 28 days, which was likely due to its strong sorption to both glass and septa. The microbiological degradation of 8:2 FTOH appears to follow multiple pathways, with neither beta-oxidation nor any identified enzyme-catalyzed reaction as a single dominant mechanism (Wang et al., 2005a; Wang et al., 2005b).

The impact of carrier solvents (ethanol, octanol, and 1,4-dioxane), which may also serve as carbon sources, on the aerobic biodegradation of 8:2 FTOHG was assessed in a clay loam (Liu et al., 2007). Biodegradation pathways were similar regardless of the solvent; however, significant differences in 8:2 FTOH degradation rates were observed: 1,4-dioxane >ethanol >octanol. In the presence of 1,4-dioxane, which is not easily biodegraded, 8:2 FTOH degradation was the fastest. With octanol, which is a structural analogue of 8:2 FTOH, the transformation was inhibited, but upon depletion of octanol, 8:2 FTOH was biodegraded (Liu et al., 2007).

Bacterial communities from sewage sludge were exposed to a mixture of perfluorinated alkylated substances (PFASs) under aerobic or anaerobic (nitrogen atmosphere) conditions (Saez et al., 2008). Compared to sterile controls, the PFASs used did not experience a significant decrease in concentration under either atmospheric condition. A decrease of 8:2 FTOH occurred in both aerobic and control systems, which may have been due to either sorption to solid matrices or bacterial activity resulting from incomplete sterilization of controls.

Fluorinated compounds other than FTOHs, such as perfluorooctane sulfonate (PFOS) and partly fluorinated non-ionic alkylpolyglycol ether (FAEO), are also considered TOrCs. PFOS and FAEO are found in the environment at soil concentrations lower than 10 mg/kg (Schroder, 2003). In experiments to investigate the fate of fluorinated surfactants reaching wastewaters, there was some evidence for the biodegradation and formation of metabolic intermediates of FAEO under aerobic and anaerobic conditions (Schroder, 2003). In a subsequent study, neither nonylphenol diethoxylate (NP2EO), PFOA, nor perfluorononanoate (PFNA) were biodegraded in activated sludge from a WWTP (Stasinakis et al., 2008).

In a study of multiple WWTPs, activated sludge in one WWTP was found to significantly increase the mass flows (formation) of PFOS, PFOA, PFNA, perfluorodecanoate (PFDA), and perfluoroundecanoate (PFUnDA) (Sinclair and Kannan, 2006). In a different plant, PFOA alone experienced an increase in formation. The generation of these TOrCs was likely due to the biodegradation of precursor compounds such as FTOHs. PFOA dominance in activated sludge decreased, while PFDA and PFUnDA concentrations increased, suggesting preferential partitioning of longer-chain PFCs to sludge as expected from mobility studies (Chapter 5.0).

6.6 Plasticizers

The majority of plasticizer biodegradation research has focused on the mechanisms and kinetics of phthalates transformation in sewage treatment plants and biosolids. However, BPA is the only plasticizer than was included within the scope of this study, and the data on BPA degradation are limited. In one study using a manometric respirometry test and activated sludge as inoculum, 4- nonylphenol (4-NP) and BPA were determined to be biodegradable (Stasinakis et al., 2008). In this same study, partial biodegradation was also observed for di-(2-ethylhexyl) phthalate (DEHP; 58.7 ± 5.7%, n = 3) and nonylphenol monoethoxylate (NP1EO) (25.9 ± 8.1%, n = 3), indicating their possible biodegradation in wastewater treatment systems. Using first order kinetics to describe biodegradation of the target compounds, the following half-lives were calculated: 4.3 ± 0.6 days for 4-NP, 1.3 ± 0.2 days for BPA, 1.8 ± 0.5 days for TCS, 6.9 ± 2.6 days for DEHP.

A Gram-negative aerobic bacterium capable of utilizing BPA as a sole carbon source was isolated from unspecified WWTP sludge (Lobos et al., 1992). BPA concentrations were approximately 1.5 mM (near saturation) and temperature and pH were held constant at 30°C and 6.5-7.0, respectively. Approximately 60% BPA was mineralized to CO_2 while the remainder was incorporated into cell biomass or soluble organic molecules.

The effects of biosolids on DEHP biodegradation was shown to vary with soil, DEHP concentration, and biosolids incubation time (Fairbanks et al., 1985). DEHP was mineralized to CO_2 with half lives ranging from 8-72 days. After 146 days, 76-93% ^{14}C-DEHP was mineralized. At 2.0 and 20.0 mg/kg DEHP was degraded more rapidly in previously conditioned soils compared to freshly amended soils. CO_2 generation was greater in freshly amended soils, suggesting greater availability of organic carbon in freshly amended soils compared to preconditioned soils.

6.7 Surfactants

Surfactants are ubiquitous in soaps and detergents, and their persistence in wastewater treatment plants leads to accumulation in biosolids. An active field of research has evaluated the ability of activated sludge to degrade surfactants. While considerable biodegradation data are available for surfactants in general, few data are available for the surfactants included within the scope of this study such as 4-cumylphenol and 4-tert-octyl phenol. Furthermore, data pertaining to surfactant degradation in soil systems are scarce. However, parallels can be drawn for related surfactants in other systems or for non-priority surfactants in biosolids-amended soils.

Sharvelle et al. (2007) studied the biodegradation kinetics of sodium lauryl ether sulfate, disodium cocoamphodiacetate (DSCADA), and polyalcohol ethoxylate in activated sludge cultured in batch experiments. Within the first 24 hours, biodegradation as measured via chemical oxygen demand removal was rapid but not complete; only 40 - 70% of the surfactant molecules were readily biodegradable. Using the Michaelis-Menten model, degradation kinetics were described: v_{max} and K_s for sodium lauryl ether sulfate were 0.21 h^{-1} and 13 mg/L; respectively; 0.092 h^{-1} and 2.9 mg/L for DSCADA; and 0.19 h^{-1} and 6.3 mg/L for polyalcohol ethoxylate.

In soil studies, 4-NP soil concentrations of 1 - 250 mg/kg were mineralized with half-lives ranging from 4.5 ± 0.5 days to 16.7 ± 2.28 days (Topp and Starratt, 2000). However, mineralization was slower in sewage sludge or heavily amended soils, which may be a result of high biochemical oxygen demand (BOD). The rate of 4-NP mineralization also decreased with decreasing temperature. Biosolids containing 900 mg/kg 4-NP were mixed with agricultural soils at 1.7 kg/m^2 in the upper 4 cm of the soil column (Brown et al., 2009). 4-NP half-lives for these experiments ranged from 16 to 23 days with faster removal found in soils that were planted with winter wheat when contrasted to unplanted soils. Attenuation was also observed for eight 4-NP isomers with minimal migration into leachates or plant leaves. Sorption of 4-NP has also been shown to affect its biodegradation in river sediments (de Weert et al., 2010). Degradation of 4-NP proceeds as the compound steadily desorbs off of sediment particles. Resuspension of 4-NP-contaminated sediment results in a sudden increase in 4-NP desorption which is subsequently biodegraded.

In a calcareous sandy soil and an acidic clayey soil treated with municipal biosolids, 4-NP degradation was biphasic (Hseu, 2006). In the calcareous soil, 4-NP was degraded rapidly within 28 days, and then degraded more completely in a second, slower phase. In the acidic clayey soil, initial degradation was slower but reached the same level of degradation over the long term incubation. 4-NP half-lives increased with increasing concentration, and ranged from 7.2 to 14 days for the calcareous soil and 7.7 to 19 days for the clayey soil (Table 6-6).

Table 6-6. Constants of 4-NP Biodegradation in the Studied Soils Treated with Biosolids.

Treatment	$t_{1/2}$ (days)	k_l (day^{-1})	r^2
Soil A (Calcareous sandy soil)			
Soil + biosolids (80 mg 4-NP / kg)	7.2 (1.5)	0.096 (0.01)	0.97
Soil + biosolids (160 mg 4-NP / kg)	12 (1.8)	0.058 (0.018)	0.95
Soil + biosolids (240 mg 4-NP / kg)	14 (1.1)	0.048 (0.012)	0.98
Soil B (Acidic clayey soil)			
Soil + biosolids (80 mg 4-NP / kg)	7.7 (1.2)	0.090 (0.021)	0.99
Soil + biosolids (160 mg 4-NP / kg)	14 (1.6)	0.049 (0.018)	0.98
Soil + biosolids (240 mg 4-NP / kg)	19 (2.0)	0.035 (0.011)	0.98

Data from (Hseu, 2006)

To gain perspective on the effect of concomitant pollutant presence, the effects of surfactants on naphthalene and phenanthrene biodegradation, and vice versa, were also investigated using activated sludge, naphthalene-acclimated organisms (NMO), and phenanthrene-acclimated organisms (PMO) (Chen and Keith A. Strevett, 2001). The anionic surfactants tested included sodium dodecyl sulfate (SDS) and sodium dodecyl benzene sulfonate (SDBS), and the two nonionic surfactants included POE (20) sorbitan monooleate (T-maz-80) and octylphenol poly(ethyleneoxy) ethanol (CA-620). In these experiments, the presence of SDS did not impact naphthalene biodegradation rates, although SDBS inhibited degradation and T-maz-80 and CA-620 reduced degradation. The presence of SDBS, CA-620 and T-maz-80 inhibited phenanthrene biodegradation, while SDS merely reduced it. The presence f naphthalene promoted CA-620 degradation, which was not degraded in its absence (Table 6-7). In the presence of phenanthrene, SDS biodegradation was reduced when compared to controls devoid of phenanthrene. In naphthalene-acclimated organisms, naphthalene was preferred over SDS. A similar trend was observed for T-maz-80.

Table 6-7. Biodegradation Kinetics of Surfactants.

Surfactant	Initial Concentration (mg/L)	Source of Biodegradation	V_{max} (h^{-1})	First-order Rate Constant, K (d^{-1})	Half saturation, K_S (mg/L)
SDS	600	Activated sludge	n/a*	1.44 ±-0.13	n/a
SDS	600	NMO		0.17 ± 0.02	n/a
SDS	600	PMO	0.37	1.68 ± 0.14	0.1
SDBS	100	Activated sludge	n/a	0.07 ± 0.01	n/a
SDBS	100	NMO	n/a	0.41 ± 0.02	n/a
SDBS	100	PMO	n/d**	n/a	n/d
T-maz-80	7.5	Activated sludge	n/a	0.38 ± 0.02	n/a
T-maz-80	7.5	NMO	0.71	0.12 ± 0.02	0.12
T-maz-80	7.5	PMO	n/a	0.12 ± 0.02	n/a
CA-620	12	Activated sludge	n/a	0.09 ± 0.02	n/a
CA-620	12	NMO	n/d	n/d	n/d
CA-620	12	PMO	n/a	0.14 ± 0.05	n/a

Data from (Chen and Keith A. Strevett, 2001)
* Data not available; ** No degradation

6.8 Conclusions

Biodegradation rates of TOrCs depend on multiple factors (Table 6-8) including the nature of the chemical (partitioning characteristics, speciation), the nature of the degrading system (soil vs. activated sludge vs. bacterial cultures, presence or absence of additional carbon sources), and environmental factors (temperature, pH, oxic conditions, chemical concentration). The extreme variability of these processes among even similar environmental systems makes broad generalizations challenging, but knowledge gained contributes to a better understanding of a TOrC's persistence in biosolids and biosolids-amended soils.

Table 6-8. Summary of Factors Affecting Degradation of TOrCs in Soil, Biosolids, and Aqueous Systems.

Compound Class	Factors Affecting Degradation (From Literature)
PPCPs	Soil type, biosolids type, oxic conditions, TOrC concentration, sorption, degrading organism, pH, temperature, metal cations
Steroidal Chemicals	Temperature, oxic conditions, degrading organism
BFRs	Speciation, degrading organism
PFCs	Additional carbon sources
Plasticizers	Soil type, TOrC concentration, biosolids incubation time
Surfactants	Soil type, BOD, additional carbon sources

This chapter examines the biodegradation, dissipation processes and kinetics of a number of important TOrCs found in biosolids, including PPCPs, steroidal chemicals, BFRs, PFCs and PFC precursors, plasticizers, and surfactants, with an emphasis on biodegradation in amended soil environments. In some cases, investigations of specific TOrCs associated with the identified chemical classes could not be found (plasticizers and surfactants) and similar chemicals in the class were selected for analogy. Similarly, some classes have benefited from research in soil systems while others lack such data; therefore, systems involving WWTP sludge or microbiological assays were occasionally considered with the hope that these systems could provide relevant, transferable data for soil systems. However, the merit of such extrapolations to the unique biogeochemical matrix presented by biosolids-amended soils is questionable at best.

As a class, PPCPs encompass a variety of compounds, including antimicrobials and the ubiquitously consumed caffeine and analgesics. Compared to other TOrC classes, data pertaining to the degradation of PPCPs in biosolids-amended soils were abundant. However, the distribution of available data among high-priority PPCPs was uneven, with the bulk of data pertaining to TCC and TCS and only a modest amount of data pertaining to tetracyclines or fluoroquinolones. The degradation of TCC and TCS has been studied in both amended and un-amended soils. In amended soils, TCC and TCS concentrations were a function of soil depth and the type of biosolids employed (Al-Rajab et al., 2009; Xia et al., 2010), while oxic conditions determined TCC and TCS persistence in un-amended soils (Ying et al., 2007). Other high-priority antibiotics such as tetracycline and doxycycline sorb strongly to biosolids in WWTP and survive storage, facilitating their persistence in post-treatment biosolids (Wu et al., 2009). The unavailable data for several high-priority PPCPs (Table 2-1) and the fact that available data are unevenly distributed among compounds indicates a continuing need for research into the degradation of PPCPs in biosolids-amended soils.

Degradation of steroidal chemicals included in this review has primarily been studied in un-amended soils and freshwater sediments. In un-amended soils, the dissipation of EE2, E2, and E1 occurred via both biotic and abiotic processes (Colucci et al., 2001; Colucci and Topp, 2001).

Microbe-mediated mineralization of these compounds comprised only a fraction of the observed dissipation from soil. Environmental factors such as temperature and soil moisture content impacted dissipation of steroidal chemicals in soil, while the reducing environment was important for biodegradation in lake sediment. While these data provide a basic understanding of the behavior of steroidal chemicals in soil systems, the effect that biosolids have on these processes remains largely unclear. Soil amendment was shown to enhance biotransformation of E2, but not testosterone, compared to un-amended soils (Jacobsen et al., 2005). Additional research is needed to determine whether these results are due to biotic or abiotic biosolids constituents. The metabolic functions of the degrading microbial community, such as nitrifying capacity, affected biodegradation in WWTP sludge suggesting a possible correlation to processes and community composition (Vader et al., 2000).

Biodegradation of BFRs and PFCs was observed primarily by microbiological assays and analysis of WWTP sludge. In biosolids-amended soil, BDE 47 and 99 were the dominant congeners in the topsoil (Arnold et al., 2008). However, in laboratory studies degradation of BFRs (primarily PBDEs) depended on chemical speciation and the degrading microbes. For example, the rate of PBDE debromination was affected by the number of Br atoms on the compound where fewer Br atoms (4-6) translated to more rapid degradation rates compared to increased bromination (7-10) (He et al., 2006). Debromination of multiple PBDEs resulted in the accumulation of toxic intermediates, such as BDE 99 and BDE 154. Similarly, the biodegradation of PFC precursors can also generate toxic byproducts, such as the transformation of 8:2 FTOH to PFOA. FTOHs are transformed via both oxidation and enzyme-mediated reactions by mixed and pure bacterial cultures in what may be a co-metabolic process. While these reports contribute to our understanding of recalcitrance and degradation products, data relating to the degradation of BFRs and PFCs in biosolids-amended soils were not found.

In activated sludge, partial degradation of certain plasticizers and a number of surfactants was observed; however, we were unable to find data that directly related to the TOrCs identified in this study either in liquid or soil systems. Biosolids-specific biodegradation data pertaining to plasticizers and surfactants were only available for compounds that have been excluded from this study (Table 2-2) such as DEHP and 4-nonylphenol. However, the data were included to better understand the degradation of similar compounds. The effects of biosolids on DEHP biodegradation varied with soil, DEHP concentration, and biosolids incubation time (Fairbanks et al., 1985). Degradation of 4-NP in amended soils is biphasic (Hseu, 2006) and is affected by temperature, BOD, and whether the soil is planted or unplanted (Topp and Starratt, 2000; Brown et al., 2009). More generally, degradation of surfactants appears to depend on the specific chemical, the degrading culture, and the presence or absence of additional carbon compounds.

The data gaps identified in this chapter clearly indicate a need for further research into select TOrCs to better understand their biodegradation and potential risks in biosolids-amended soil. Extrapolation of aqueous phase biodegradation data to biosolids-amended soils is fraught with uncertainty and further complicated by factors such as soil heterogeneities or climate. Some soil data are available for PPCPs and steroidal chemicals, but the data are limited to a subset of the priority TOrCs identified in Table 2-1. Data pertaining to the biodegradation of BFRs, PFC precursors, and surfactants were obtained primarily via analysis of bacterial cultures and WWTP sludge. Data for biosolids or biosolids-amended soils were essentially non-existent. These compounds would benefit most from additional biosolids-focused research. It is unclear whether microbial community composition or the presence of certain microorganisms can be linked to biodegradation potential. Furthermore, the impact of TOrC mixtures to assess the impact of one priority TOrC on the behavior of another needs further exploration. To better understand the persistence of TOrCs in the complex matrix of biosolids-amended soils, future research should ideally focus on long-term field studies and studies involving the biodegradation of chemical mixtures.

Table 6-9 provides a summary of the general data availability with respect to the high priority TOrCs included in this study. The decision used to bin the TOrC classes or subclasses are provided, though in some cases, expert judgment was required to consider data that did not readily fit within the framework developed. As is clear from the table, significant data gaps with respect to persistence are evident for many of the TOrCs included in this study.

Table 6-9. Summary of Persistence Data Availability for the High Priority TOrCs.

Chemical Class	Data Availability
BFRs	Tier 1
PFCs and PFC Precursors	Tier 1
PPCPs: Antimicrobial Agents	Tier 3
PPCPs: Antibiotics	Tier 1
PPCPs: Synthetic Musks	Tier 3
PPCPs: Other	Tier 0
Plasticizers	Tier 1
Steroidal Chemicals	Tier 2
Surfactants	Tier 0

Data Availability Ranking Decision Criteria:

Tier 0 (No Data)	Essentially no data were available of this type for this class or subclass of TOrCs, including data that could be used for modeling.
Tier 1	For the majority of TOrCs in this class or subclass, biodegradation studies have been conducted (disregarding of incubation medium or environment).
Tier 2	For the majority of TOrCs in this class or subclass, biodegradation studies have been conducted in soils.
Tier 3	For the majority of TOrCs in this class or subclass, field-scale persistence studies have been conducted.

CHAPTER 7.0

BIOAVAILABILITY AND BIOACCUMULATION OF BIOSOLIDS-BORNE TRACE ORGANIC CHEMICALS IN SOILS

7.1 Introduction

Once introduced to the environment, trace organic chemicals (TOrCs) have the potential to impact biota. How much impact a chemical will have on biota depends on its bioavailability. Bioavailability and bioaccumulation are particularly important for biosolids-amended soils as once lower trophic level species such as worms accumulate TOrCs, there is the potential for transfer of the TOrC to higher trophic levels. Bioaccumulation in plants may also create an additional exposure route for any organisms consuming the plants (including humans). Thus, accumulation in plants and animals is important for an accurate assessment of the risks associated with biosolids-borne TOrCs.

Where data are available, bioaccumulation in plant species is discussed in this chapter. However, data on accumulation in plants are relatively limited for the targeted TOrCs. As discussed in two reviews of this topic (O'Connor, 1996; Katayama et al., 2010), bioaccumulation in plants can be both passive and active, with the former being much easier to predict from chemical-specific and soil-specific physicochemical parameters. More data are available regarding bioaccumulation in animals; however data are still lacking for many of the TOrCs targeted in this study.

The focus of the data summarized in this chapter is on the initial bioaccumulation of TOrCs from biosolids amended soils, either to plants or animals, and does not specifically address trophic transfer of the targeted TOrCs up the foodchain. Depending on the TOrC and the foodchain, *biomagnification* of the TOrC may also occur, a process by which the body burden of the TOrC increases as the trophic level increases. While no studies were identified detailing the biomagnification of TOrCs from biosolids-amended soils, consideration of such processes may be important when modeling the risks associated with TOrCs in biosolids-amended soils. For example, a recent study conducted for triclocarban (TCC) suggested the most sensitive pathway for adverse effects of TCC in biosolids-amended soils arose due to the potential for trophic transfer of TCC from earthworms to birds (Snyder, 2009). Unfortunately, laboratory and/or field-based data addressing this process for the targeted TOrCs are extremely limited.

Lastly, though a detailed discussion of the various factors impacting the bioavailability of TOrCs in soils is beyond the scope of this effort, it is important to note that, in general, the more strongly bound the TOrC is to the soil, the less bioavailable it is and the less likely it is to bioaccumulate. An extensive and excellent review of the various processes affecting bioaccumulation and modeling approaches used to predict bioaccumulation of TOrCs from soils into both plants and animals has recently been published (Katayama et al., 2010). Unfortunately, the complexity of the biosolids matrix, including the potential for TOrCs to undergo irreversible sorption (or reversible sorption with much slower desorption kinetics) is not always captured with the conventional modeling approaches. Moreover, the traditional hydrophobic organic contaminant partitioning paradigms may not apply to some of the targeted TOrCs, particularly the perfluorochemicals (PFCs) and their precursors. For these reasons, experimental data, particularly from exposures conducted in biosolids-amended soils is of the utmost importance and was the target of the data gap assessment.

7.1.1 Bioaccumulation Metrics

The chapter primarily discusses two measurements of the bioavailability of a chemical, both of which are commonly used to evaluate bioaccumulation in soil- and sediment-dwelling organisms such as oligochaetes. The first parameter is the bioaccumulation factor (BAF). The BAF takes into account all routes of exposure from medium to organism (e.g., diet, dermal, respiratory tissue) and is the ratio of the concentration in biota to the concentration in the exposure medium, assuming steady state. Units for BAF values are variously expressed, depending on the exposure medium and whether the concentration in the organism is expressed on a wet weight (ww), dry weight (dw), or lipid-normalized basis (lip). For example, if the concentration of TOrC in the organism is expressed on a wet weight basis and the concentration of the TOrC in the exposure medium is expressed on a dry weight basis (i.e., for a soil), the units for resulting BAF would be kg_{dw}/kg_{ww}:

$$BAF = \frac{C_{org,ww}\left(mg/kg_{ww}\right)}{C_{soil,dw}\left(mg/kg_{dw}\right)} \qquad 7.1.1\text{-}1$$

The standard convention is to express BAFs on a kg_{dw}/kg_{dw} basis, implying a "dimensionless" BAF, though this can be misleading. To ensure clarity, we have included the units whenever available. Similar to BAFs, bioconcentration factors (BCFs) are the steady-state ratio of chemical residue in the organism to chemical concentration in the water *only*. BCF values are used in expressing chemical accumulation for organisms that do not ingest food (e.g., algae) and in laboratory tests where organisms are not fed and no sediment is present. BCFs may also be expressed on a wet weight (ww), dry weight (dw), or lipid-normalized basis (lip). Thus, when the organism concentration is expressed on a wet weight basis and the aqueous concentration is expressed on a volume basis (i.e., mg/L), the units for BCFs are L/kg_{ww}.

The second parameter is the biota-soil accumulation factor (BSAF), sometimes also referred to as a biota-sediment accumulation factor. A BSAF is a ratio of the concentration in biota to the concentration in soil (or sediment), and is typically normalized to the lipid content of the organism and the organic carbon content of the soil or sediment (resulting in units of kg_{oc}/kg_{lip}). In essence, a BSAF is a specific type of BAF that is particularly relevant for soils. The rationale for the normalization to organism lipids and soil organic carbon is that the net transfer of a TOrC between these two pools of organic carbon is assumed to be zero: the system is at steady state (Wong et al., 2001). Deviations from the normal range of BSAF values (1-2) (Wong et al., 2001) suggests other factors are affecting the bioavailability of the chemicals. This is particularly important for biosolids, as incorporation of the biosolids-borne TOrCs into the soil organic matter may result in significant reductions in bioavailability. Unfortunately, as has been observed when trying to model the bioaccumulation of many TOrCs in fish (Arnot and Gobas,

2006), the biological activity of many of the TOrCs (either in the form of specific metabolic transformations or affinities for specific tissues) creates additional complications that are not easily addressed using the current BSAF model. Nevertheless, BSAFs can provide useful metrics to measure the relative bioavailability of a TOrC under specific conditions.

The issue of steady-state is particularly relevant for laboratory-based experiments. Without data on the uptake of specific TOrCs over time (i.e., uptake kinetics), it is difficult to assess whether a laboratory system has truly reached steady-state. For example, though 28 days is often considered sufficient for steady-state to be reach for chemical uptake into the sediment-dwelling oligochaete *Lumbriculus variegatus* (U.S. EPA, 2000), uptake kinetic data for PFCs in these organisms indicate steady-state was not immediately evident even after 56 days of exposure (Higgins et al., 2007). If appropriate uptake kinetic data are collected, the steady-state body burdens can be estimated even if steady-state has not been obtained. Such extrapolations are significantly more difficult for plant uptake experiments, where growth dilution significantly complicates the extrapolations.

7.2 Pharmaceuticals and Personal Care Products

7.2.1 Antimicrobial Agents

Bioaccumulation in Plants

Though not directly relevant to biosolids-amended soils, the bioconcentration of the antimicrobial agents triclosan (TCS) and TCC has been studied in algae exposed in a stream near a wastewater treatment plant (WWTP) outfall (Coogan and La Point, 2008). After two weeks of exposure, mean TCS and TCC concentrations in algae were 162 ng/g_{ww} and 367 ng/g_{ww}, respectively. Calculated BCF values for TCC and TCS in algae were 1,900 L/kg_{ww} and 1,400 L/kg_{ww}, respectively (Coogan and La Point, 2008). In a more relevant study, the uptake of TCC in Bahia grass was examined (Snyder, 2009). The Bahia grass was grown in biosolids-amended soil, treated with one of eleven different biosolids so TCC exposure concentrations varied. The grass was harvested every 4-8 weeks for more than a year. TCC concentrations in grass clippings from early harvests (expected to maximize uptake) ranged from 0.01 - 1.2 ng/g_{dw}. Calculated BAF values ranged from 0.00041 - 0.008. The author notes that BAF values of this magnitude are generally considered to be insignificant (Snyder, 2009).

Bioaccumulation in Animals

Studies were identified that investigated the bioaccumulation of TCC and TCS in earthworms, the aquatic snail *Helisoma trivolvis*, and the aquatic oligochaete *Lumbriculus variegatus* (Coogan and La Point, 2008; Kinney et al., 2008; Higgins et al., 2009; Snyder, 2009). Bioaccumulation in field-collected earthworms of a large suite of TOrCs commonly found in biosolids was studied (Kinney et al., 2008). Concentrations of TCS in worms collected from fields amended with biosolids ranged from 1.740 - 2,610 μg/g_{dw} (Kinney et al., 2008). The data were used to calculate a BAF range of 10.8 - 27 kg_{dw}/kg_{dw} for TCS in earthworms, as summarized in Table 7-1 (Kinney et al., 2008). A second study also looked at bioaccumulation of TCC in earthworms (Snyder, 2009). Soil was amended with biosolids spiked with an additional 70 - 700 mg/kg TCC and earthworms were exposed to the biosolids-amended soils for four weeks. Surviving worms had TCC concentrations of 36.5 - 127 mg/kg_{dw} resulting in BAF values of 2.2 - 18 (Snyder, 2009).

After two weeks of exposure in a stream near a WWTP outfall, mean concentrations of TCS and TCC in *H. trivolvis* were 58.7 ng/g_{ww} and 299 ng/g_{ww}, respectively (Coogan and La Point, 2008). Calculated BAF values for TCC and TCS in *H. trivolvis* were 1,600 and 500 L/kg_{ww}, respectively, as summarized in Table 7-1. BAFs for TCC were about three times greater than those for TCS (Coogan and La Point, 2008).

Bioaccumulation of TCC in *L. variegatus* from spiked sediments resulted in maximum concentrations of 42 µg/g_{ww} when exposed to 22.4 ± 7.6 µg/g_{dw} in the sediment (Higgins et al., 2009). Maximum concentrations were observed on day five of a 56-day study. TCC concentrations in *L. variegatus* increased rapidly through day five and then decreased through the end of the 56-day period, though no decrease in sediment TCC levels was observed. BSAF and BAF values calculated for TCC in this study were 1.6-2.2 (kg_{oc}/kg_{lip}) and 1,600-2,200 (L/kg_{ww}), respectively (Higgins et al., 2009).

Factors Affecting Bioavailability

Studies of bioaccumulation of TCC in the aquatic worm *L. variegatus* showed an initial rapid increase of TCC concentrations in the first five days followed by a decrease in body burden through the end of the 56-day study (Higgins et al., 2009). The authors identify several potential causes of this trend including decreasing TCC concentrations in sediment, a decrease in the amount of bioavailable TCC in sediment, and the possible transformation of TCC within *L. variegatus*. However, the capability of *L. variegatus* to metabolize persistent hydrophobic organic compounds is known to be limited and there was no evidence TCC transformation (Higgins et al., 2009). Additionally, the measured TCC concentration in sediment did not decline over time, so it is likely that changes in the bioavailable fraction of TCC over time caused the observed decrease in the body burden of TCC (Higgins et al., 2009).

7.2.2 Tetracycline Antibiotics

Two studies were found regarding the bioaccumulation of tetracycline antibiotics in plants (Kumar et al., 2005; Kong et al., 2007). The studies looked at uptake of chlortetracycline (CTC) and oxytetracycline (OTC), which are chemicals that were not included as part of this review. However, the research is summarized here because of the structural similarities among the tetracycline antibiotics.

CTC has been found to accumulate in green onion, cabbage, and corn from both artificially spiked soil and from soil treated with swine manure (Kumar et al., 2005). The chemical accumulated in green onions and cabbage from artificially spiked soil (corn was not tested) and in green onions, cabbage, and corn from soil treated with swine manure. Analysis of the shoot portions of the plants found concentrations of CTC of 2 - 17 mg/g_{ww}. Unfortunately, exposure concentrations were not readily available in the report. It appears that artificially spiked soil was treated with a 20 µg/L solution of CTC, and spiked manure was treated with an additional 100 mg/kg of CTC. However, exposure concentrations were not provided. Concentrations in plants increased with increasing concentrations of CTC in both the artificially spiked soil and in soil treated with manure; however, concentrations in both cases decreased with time. The authors attributed the decrease to an increase in plant biomass (growth dilution) and/or a decrease in the bioavailable portion of CTC in the soil (Kumar et al., 2005).

Table 7-1. Bioaccumulation Parameters for the Selected TOrCs in Animals.

Chemical	Chemical Class	BAF (units)	BSAF (kg_{oc}/kg_{lip})	Organism	Data Sources
Galaxolide (HHCB)	PPCPs	0.05 - 3.1 (kg_{dw}/kg_{dw}) [a]		Earthworm	1
Tonalide (AHTN)	PPCPs	0.1 - 1 (kg_{dw}/kg_{dw}) [a]		Earthworm	1
Triclocarban (TCC)	PPCPs	1600 (L/kg_{ww})		*H. trivolvis*	
		1900 (L/kg_{ww})		*Cladophora*	
		1600 - 2200 (L/kg_{ww})	1.6 - 2.2	*L. variegatus*	
		0.00041-0.008		Bahia Grass	
		2.2 - 18		Earthworm	2, 3, 4
Triclosan (TCS)	PPCPs	10.8 - 27 (kg_{dw}/kg_{dw}) [a]		Earthworm	
		500 (L/kg_{ww})		*H. trivolvis*	
		1400 (L/kg_{ww})		*Cladophora*	1, 3
17α-ethynylestradiol (EE2)	Steroidal Chemicals		190	*L. variegatus*	5
BDE 47	BFRs			*L. variegates*	
			1 - 9[b]	Earthworm[d]	
		1.7 - 8.1 (kg_{dw}/kg_{dw})	2.5[c]	Earthworm[d]	6, 7, 8
BDE 66	BFRs		2 - 8.5[b]	Earthworm[d]	7
BDE 85	BFRs	4.9 (kg_{dw}/kg_{dw})		*L. variegatus*	6
BDE 99	BFRs	0.8 - 4.0 (kg_{dw}/kg_{dw})		*L. variegates*	
			0.5-5.5[b]	Earthworm[d]	
			2.1[c]	Earthworm[d]	6, 7, 8
BDE 100	BFRs	1.2 - 9.9 (kg_{dw}/kg_{dw})		*L. variegates*	
			1.5 - 17[b]	Earthworm[d]	
			2.3[c]	Earthworm[d]	6, 7, 8
BDE 126	BFRs		0.45 - 1.5[b]	Earthworm[d]	7
BDE 138	BFRs		0.5 - 4.5[b]	Earthworm[d]	7
BDE 153	BFRs	4.7 (kg_{dw}/kg_{dw})		*L. variegatus*	
			0.5 - 3[b]	Earthworm[d]	
			1.25[c]	Earthworm[d]	6, 7, 8
BDE 154	BFRs	9.1 (kg_{dw}/kg_{dw})		*L. variegates*	
			1.15[c]	Earthworm[d]	6, 8
BDE 196	BFRs		0.3[c]	Earthworm[d]	8
BDE 197	BFRs		1[c]	Earthworm[d]	8
BDE 206	BFRs		0.15[c]	Earthworm[d]	8
BDE 207	BFRs		0.5[c]	Earthworm[d]	8
BDE 209	BFRs		0.15[c]	Earthworm[d]	8
PFOA	PFCs and Precursors		33	*L. variegatus*	9
PFNA	PFCs and Precursors		55	*L. variegatus*	9
PFDA	PFCs and Precursors		35	*L. variegatus*	9
PFUnDA	PFCs and Precursors		21	*L. variegatus*	9
N-EtFOSAA	PFCs and Precursors		7	*L. variegatus*	9
PFOS	PFCs and Precursors		42	*L. variegatus*	9
PFDS	PFCs and Precursors		17	*L. variegatus*	9

Data sources:

1 (Kinney et al., 2008)
2 (Higgins et al., 2009)
3 (Coogan and La Point, 2008)
4 (Snyder, 2009)
5 (Liebig et al., 2005)
6 (Ciparis and Hale, 2005)
7 (Matscheko et al., 2002)
8 (Sellstrom et al., 2005)
9 (Higgins et al., 2007)

[a] Units for BAF values are assumed to be in kg_{dw}/kg_{dw}; however, units of soil concentration were not explicitly stated in the paper.

[b] BSAF values are assumed to be normalized to organic matter which is approximately equal to two times f_{oc}. Reported BSAF values were multiplied by 0.5 to convert to units of kg_{oc}/kg_{lip}.

[c] BSAF values were originally normalized to ignition loss which is assumed to represent organic matter. Organic matter is approximately equal to two times f_{oc}, so reported BSAF values were multiplied by 0.5 to convert to units of kg_{oc}/kg_{lip}

[d] Species of earthworms include *Lumbricus terrestris, Lumbricus* spp., *Aporrectodea caliginosa, A. rosea,* and *Allolobophora chlorotica.*

Accumulation studies involving hydroponic systems (plants grown only in water) are of limited utility with respect to understanding accumulation in plants from biosolids-amended soils, but can serve as "worse case' scenarios in which the TOrC is present in its most bioavailable form. In one such study, alfalfa accumulated the tetracycline antibiotic OTC from a nutrient solution spiked with the chemical (Kong et al., 2007). Bioaccumulation depended on the pH of the exposure solution; greater bioaccumulation occurred at pH 7 than at pH 5. Alfalfa plants treated with OTC exhibited negative effects such as yellowing of leaves (Kong et al., 2007). After 10 hrs of exposure to a concentration 0.02 mM, concentrations of OTC in alfalfa were approximately 350 nmol/g fresh weight (Kong et al., 2007).

7.2.3 Fluoroquinolone Antibiotics

One study was found which touched briefly on the bioaccumulation of fluoroquinolones in lettuce and carrots (Boxall et al., 2006). The fluoroquinolone enrofloxacin accumulated in carrot roots but not in lettuce. Enrofloxacin is mentioned here due to the structural similarities between this chemical and other fluoroquinolones such as ciprofloxacin (CIP) and ofloxacin, which are targeted TOrCs.

7.2.4 Synthetic Musks

Bioaccumulation in Plants

Studies of bioaccumulation of acetyl-hexamethyl-tetrahydronaphthalene (AHTN) and hexahydro hexamethylcyclopentabenzopyran (HHCB) in lettuce and carrots from a sewage sludge-soil mixture have been completed (Litz et al., 2007). All plants were grown in a sewage sludge-soil mixture containing 30 mg/kg of either HHCB or AHTN. AHTN and HHCB accumulated in both lettuce and carrots, but bioaccumulation was much greater in carrots. In lettuce, the maximum concentrations of HHCB and AHTN were 290 $\mu g/kg_{dw}$ and 820 $\mu g/kg_{dw}$, respectively. In carrot roots, concentrations of HHCB and AHTN were much higher at 21 mg/kg_{dw} and 18 mg/kg_{dw}, respectively. Concentrations of both chemicals decreased in both lettuce and carrots over time starting at week 6 through the end of the experiment at week 12. There was no degradation of HHCB or AHTN during the exposure period. Bioaccumulation of AHTN and HHCB in carrots decreased with time, and was attributed to growth dilution (Litz et al., 2007).

Bioaccumulation in Animals

Bioaccumulation in earthworms of a large suite of chemicals commonly found in biosolids was studied in a field setting (Kinney et al., 2008). The suite of chemicals included HHCB and AHTN. Because the earthworms were collected in a field setting, dry weight exposure concentrations varied from 633-2,770 μg/kg for HHCB and 113 - 773 μg/kg for AHTN. Dry weight concentrations in earthworms collected from biosolids-amended soils were 49 - 3340 μg/kg for HHCB and 19 - 279 μg/kg for AHTN. Calculated BAF values for the chemicals ranged from 0.05 - 3.1 kg_{dw}/kg_{dw} for HHCB and 0.1 - 1 kg_{dw}/kg_{dw} for AHTN (Table 7-1).

7.3 Steroidal Chemicals

Bioaccumulation of radiolabeled 17α-ethinyl estradiol (EE2) in the aquatic worm *L. variegatus* resulted in a near-linear increase in concentrations over a 35-day test period (Liebig et al., 2005). *L. variegatus* was exposed to EE2 via spiked, artificial sediments. At the end of the 35-day exposure period, the measured BSAF was 75 kg_{oc}/kg_{lip}; however, steady was not obtained. The calculated steady state BSAF is 191 kg_{oc}/kg_{lip} (Liebig et al., 2005).

7.4 Brominated Flame Retardants

Bioaccumulation in Plants

The bioaccumulation of polybrominated biphenyl ethers (PBDEs) has been studied in various plants including radish, zucchini, and the aquatic plant *Ceratophyllum demersum* (Mueller et al., 2006; Sun et al., 2008a). Radish and zucchini accumulated penta BDEs, including BDE 47, BDE 99, and BDE 100 in concentrations as high as 4 $\mu g/kg_{dw}$ in plant tissue exposed to soils spiked to contain 75 μg/kg of total penta BDEs (Mueller et al., 2006). Of the congeners studied, BDE 100 was found to have the highest uptake in plant roots despite a small contribution to the overall PBDE concentration in the soil. In this study, zucchini roots were found to be twice as effective as radish roots in the uptake of PBDEs. In addition, zucchini exhibited greater ability than radish to translocate PBDEs to plant shoots (Mueller et al., 2006).

Only one study was identified that looked at the bioaccumulation of the flame retardant tetrabromobisphenol A (TBBPA). The aquatic plant *C. demersum* was shown to accumulate TBBPA from water (Sun et al., 2008a). Maximum concentrations of *C. demersum* in this study reached approximately 0.7 mg/g_{dw} after 14 days of exposure to 0.5 mg/L TBBPA. Bioaccumulation in *C. demersum* also increased with increasing TBBPA concentration. Concentrations of TBBPA in *C. demersum* ranged from 0 -1.2 mg/g dry weight after exposure to TBBPA concentrations ranging from 0 - 1 mg/L. This study also concluded that TBBPA might induce oxidative stress on *C. demersum*. For example, TBBPA uptake was associated with enhanced lipid peroxidation and a decline in chlorophyll content (Sun et al., 2008a).

Bioaccumulation in Animals

Studies were identified that investigated the bioaccumulation and bioavailability of PBDEs in both terrestrial and aquatic species (Hale et al., 2002; Matscheko et al., 2002; Ciparis and Hale, 2005; Sellstrom et al., 2005). Bioavailability studies of PBDEs in terrestrial species include frogs, crickets, and earthworms. Penta BDEs, including BDE 47, BDE 99, BDE 100, BDE 153, and BDE 154 accumulated in frogs and crickets housed with PBDE-bearing polyurethane foam (Hale et al., 2002). The study suggested that frogs likely accumulated the penta BDEs by consuming crickets, which directly consumed the PBDE-bearing foam. Concentrations of 10.1 mg/kg_{ww} and 14.4 mg/kg_{ww} were detected in frogs and crickets, respectively. The authors noted that neither frogs nor crickets were depurated prior to analysis, therefore detected PBDE concentrations may be elevated due to residual foam present in the digestive tracts of frogs and crickets (Hale et al., 2002).

PBDEs accumulate in earthworms from biosolids-treated soil and the information used to calculate BSAFs for some of the PBDE congeners (Matscheko et al., 2002; Sellstrom et al., 2005). The data are summarized in Table 7-1. The sum of all PBDEs (BDE 35, BDE 47, BDE 49, BDE 66, BDE 85, BDE 99, BDE 100, BDE 153, BDE 154, BDE 183, BDE 196, BDE 198/203, BDE 206, BDE 207, and BDE 209) measured in one study found concentrations ranging from 3.1 to 38,000 ng/g_{lip} in earthworms (Sellstrom et al., 2005). Additionally, higher concentrations of a particular PBDE congener in soil correlated with a higher concentration of that congener in the earthworms (Sellstrom et al., 2005). In the same study, several PBDE congeners (BDE 25, BDE 49, BDE 183, BDE 196, BDE 197, BDE 198/203, and BDE 207) were found in worms despite being found less frequently or not at all in soils. This could indicate bioconcentration of congeners that were below detection limits in soil, or it could be that earthworms metabolize higher brominated PDBEs to lower brominated congeners (Sellstrom et al., 2005).

The aquatic worm *L. variegatus* bioaccumulated PBDEs from biosolids, as well as from artificial sediment spiked with penta and deca BDEs (Ciparis and Hale, 2005). The information was used to calculate BAF for applicable PBDE congeners summarized in Table 7-1. Both substrates lead to bioaccumulation of PBDEs in *L. variegatus*; however, accumulation from the artificially spiked sediment was approximately 5-10 times greater than that from biosolids. In both substrates, BDE 47 and BDE 99 were the most accumulative and bioaccumulation of BDE 209 was negligible (Ciparis and Hale, 2005).

Factors Affecting Bioavailability

The studies summarized above identified a number of factors that may impact bioavailability of PBDEs. Accumulation of PBDEs by *L. variegatus* was higher from spiked, artificial sediment than from biosolids. A variety of factors may have lead to this difference including organic matter composition, length of time the substrate was exposed to PBDEs, partitioning changes of PBDEs in the matrix, the source of the PBDEs to the matrix, and organism physiology (Ciparis and Hale, 2005).

The same study found that BDE 47 and BDE 99 were the most accumulative of the congeners studied (Table 7-1). Of the two, BDE 47 accumulated the most in *L. variegatus*. The authors identified two potential reasons for this. First, they noted that BDE 99 might have a higher depuration rate than BDE 47. The second potential cause identified was that BDE 99 has a higher octanol-water partitioning coefficient (K_{ow}) value than BDE 47, which may cause it to be strongly sorbed and less available to organisms (Ciparis and Hale, 2005). A subsequent study identified that BSAFs for PBDEs decreased with increasing K_{ow} (Sellstrom et al., 2005).

It is also possible that bioaccumulation of PDBEs is related to the degree of bromination, but some evidence shows that bioaccumulation is related more to the substitution pattern (Ciparis and Hale, 2005). For example, BAFs for BDE 47, BDE 100, and BDE 154 were higher than those for BDE 99, BDE 85, and BDE 153 though both groups contain penta and hexa BDEs. However, BDE 47, BDE 100, and BDE 154 all have a bromine in the ortho position (Ciparis and Hale, 2005).

Translocation of PBDEs from the roots to the sprouts of plants may also be congener-specific. In the bioaccumulation of PBDEs by zucchini, translocation to plant shoots increased with increasing bromine while initial uptake of PBDE congeners by plant roots was relatively consistent (Mueller et al., 2006). If translocation were purely dependent on the movement of chemicals during transpiration, lower brominated PBDEs with higher aqueous solubility (S_w) would be expected to demonstrate higher translocation. However, this was not the case; BDE 100 was translocated more than less brominated congeners. The trend is likely due to stereochemical constraints or plant physiology (Mueller et al., 2006).

7.5 Perfluorochemicals and Perfluorochemical Precursors

There is evidence for the bioaccumulation of PFCs in invertebrates. *L. variegatus* accumulated perfluorooctanoate (PFOA), perfluorononanoate (PFNA), perfluorodecanoate (PFDA), perfluoroundecanoate (PFUnDA), perfluorododecanoate (PFDoA), perfluorooctane sulfonate (PFOS), and perfluorodecane sulfonate (PFDS) from spiked sediment as well as from field sediment collected downstream of a wastewater treatment facility (Higgins et al., 2007). In addition, *L. variegatus* was found to accumulate the PFOS precursor 2-*N*-ethylperfluorooctane-sulfanamido acetic acid (*N*-EtFOSAA). Concentrations of individual PFCs in *L. variegatus* increased over the 56-day exposure period reaching maximum values between 65 and 270 ng/g_{ww} by the end of the study when exposed to 7 to 17 ng/g_{dw} PFCs in the sediment, though steady state might not have been achieved. Lipid-normalized BSAF values for PFCs ranged from 17 to 55 kg_{oc}/kg_{lip} in laboratory-spiked sediment and 4 - 177 kg_{oc}/kg_{lip} in field-collected sediment. However, lipid-normalization is inappropriate for PFCs, as they do not accumulate in lipids or adipose tissue (Higgins et al., 2007). Instead, non-lipid (but organic carbon) normalized BSAFs were calculated to be 0.22 to 1.60 kg_{oc}/kg_{ww}, for laboratory-spiked sediment and 0.02 to 0.83 kg_{oc}/kg_{ww} for field-contaminated sediment. Bioaccumulation generally decreased with increasing perfluorocarbon chain length (Higgins et al., 2007). BSAF values from this study are summarized in Table 7-1. Concentrations of *N*-EtFOSAA peaked at day 5 of the exposure period, decreased through day 28, and reached an approximate steady state thereafter. A BSAF value of 7 kg_{oc}/kg_{lip} (0.02 kg_{oc}/kg_{ww}) was reported for N-EtFOSAA (Higgins et al., 2007).

Factors Affecting Bioavailability

As discussed above, bioaccumulation studies of PFCs in *L. variegatus* decreased with increasing chain length. The reason for this trend was unclear but may have been caused by slower elimination rates in PFCs with shorter chain lengths (Higgins et al., 2007). In addition, bioaccumulation of *N*-EtFOSAA in *L. variegatus* increased until day 5 after which it decreased until day 28 and reached an approximate steady state. Several factors were suggested to play a role in this trend. First, excretion of the chemical may contribute to the trend. Second, Phase II metabolic conjugation of *N*-EtFOSAA by *L. variegatus* may facilitate excretion of the chemical. Finally, data from the study suggest that *L. variegatus* may biotransform *N*-EtFOSAA to perfluorooctane sulfonamide (FOSA) and PFOS. Notably, PFOS was observed to bioaccumulate in worms only exposed to N-EtFOSAA (Higgins et al., 2007).

7.6 Plasticizers and Surfactants

Bioaccumulation in earthworms of a large suite of chemicals commonly found in biosolids was studied (Kinney et al., 2008). While the suite of TOrCs analyzed in this study included BPA, BPA was not detected in earthworms collected from biosolids-amended soils (Kinney et al., 2008). The suite of TOrCs also included the surfactants 4-tert-octylphenol and 4-cumylphenol. Dry weight concentrations measured in earthworms collected from biosolids-amended soils were 186 - 570 µg/g, and 37 µg/g for 4-tert-octylphenol and 4-cumylphenol, respectively. However, BAF values could not be calculated for the chemicals because concentrations of the chemicals in the biosolids-amended soils were below the limits of detection (Kinney et al., 2008).

7.7 Conclusions

The land application of biosolids may introduce TOrCs to the environment that subsequently become available for bioaccumulation in plants and animals. The objective of this chapter was to summarize available information regarding the bioaccumulation of TOrCs in plants and animals. Few studies are available examining bioaccumulation and bioavailability specifically in biosolids-amended soils, so the chapter also summarizes studies of bioavailability in soils, sediments, and in some cases, aquatic environments. The species in which TOrC bioaccumulation was studied exposure concentrations, concentration in the organism, and reported BAF or BSAF values are noted when available. Certain information gaps and research needed to fully understand risks associated with land application of biosolids-borne TOrCs are apparent.

Studies that investigated the bioaccumulation of TOrCs in plants were limited to the PPCP and brominated flame retardant (BFR) compound classes. For the PPCPs, studies were available for antimicrobial agents, tetracycline antibiotics, fluoroquinolones, and synthetic musks. These chemicals were shown to accumulate in a variety of plants including grass, green onions, cabbage, corn, lettuce, and carrots. Of the BFRs , penta BDEs and TBBPA specifically accumulate in various plants including radish, zucchini, and aquatic plants. No studies were identified for several compound classes, including steroidal chemicals, PFCs, plasticizers, and surfactants.

More information was available regarding the bioaccumulation and bioavailability of TOrCs in animals, including studies of the following compound classes: pharmaceuticals and personal care products (PPCPs), steroidal chemicals, BFRs, PFCs, plasticizers, and surfactants. In the PPCP compound class, antimicrobial agents and synthetic musks were shown to accumulate in earthworms and aquatic oligochaetes. The steroidal chemical EE2 accumulates in aquatic oligochaetes. Among the BFRs, PBDEs can bioaccumulate in frogs, crickets, and aquatic oligochaetes. Similarly, chemicals in the PFC class were shown to accumulate in aquatic oligochaetes. Bioaccumulation of the plasticizer BPA was studied, but the chemical was not found in earthworms. The surfactants 4-tert-octylphenol and 4-cumylphenol, however, were found to accumulate in earthworms.

Table 7-2. Summary of Bioaccumulation Data Availability for the High Priority TOrCs.

Chemical Class	Data Availability
BFRs	Tier 2
PFCs and PFC Precursors	Tier 0
PPCPs: Antimicrobials	Tier 1
PPCPs: Antibiotics	Tier 0
PPCPs: Synthetic Musks	Tier 2
PPCPs: Other	Tier 0
Plasticizers	Tier 1
Steroidal Chemicals	Tier 1
Surfactants	Tier 1

Data Availability Ranking Decision Criteria:

Tier 0 (No Data)	Essentially no data were available of this type for this class or subclass of TOrCs, including data that could be used for modeling.
Tier 1	For the majority of TOrCs in this class or subclass, physicochemical parameters have been measured (i.e., K_{ow}) that would enable bioaccumulation potential to be assessed in both plants and animals.
Tier 2	For the majority of TOrCs in this class or subclass, bioaccumulation studies have been conducted for either plants or animals in spiked soil or sediment systems using appropriate species and analytical protocols.
Tier 3	For the majority of TOrCs in this class or subclass, realistic field-scale monitoring studies have been conducted evaluating bioaccumulation in both plants and animals from biosolids-amended soils.

Table 7-2 provides a summary of the general data availability with respect to the high priority TOrCs included in this study. The decision used to bin the TOrC classes or subclasses are provided, though in some cases, expert judgment was required to consider data that did not readily fit within the framework provided. As is clear from the table, significant data gaps with respect to bioaccumulation are evident for many of the TOrCs included in this study.

Katayama et al. (2010) reviewed the many factors can impact the bioavailability of TOrCs to plants and animals. As the chemical is accumulated, the amount present in soil will decrease. Other factors may also lead to a decrease in the exposure concentration, such as leaching and biodegradation. Even if the overall decrease in the chemical is minimal, the fraction of the chemical that is bioavailable may decrease. For instance, ongoing changes in partitioning to the solid phase (including non-reversible sorption) may decrease the fraction available to organisms (i.e., bioavailability). Thus, soil parameters affecting sorption, such as organic carbon content and soil pH, may also impact the bioavailability of a chemical. Unfortunately, many of the studies identified did not provide significant detail as to the exposure conditions. Bioaccumulation can also be affected by changes in the organism in question. Initially high bioaccumulation may be reduced as plant biomass increases, and organisms may biotransform the chemical or conjugate chemicals to facilitate their excretion.

Because of the factors discussed above and variability in research methods, it is very difficult to make comparisons of bioavailability among the various studies. This is true even for studies examining the same chemical and even more so for comparisons at the level of compound class. Parameters such as BAF and BSAF values are meant to facilitate such comparisons. Many of the studies discussed in this chapter calculated BAFs or BSAFs, summarized in Table 7-1. However, the values they do not easily lend themselves to comparison. Units are not always provided and are not always consistent, making comparisons difficult. Additionally the applicability of the BSAF values must be investigated if they are lipid-normalized. Some chemicals such as PFCs do not accumulate in lipids, rendering lipid normalization problematic, while the affinity of other TOrCs for the solid phase (i.e., tetracyclines) is not necessarily dependent on organic carbon, rendering organic carbon normalization problematic. These factors point to a need for consistency in measuring and reporting data to facilitate comparisons among the TOrCs. Factors that should be considered include the organisms used (i.e., standard organisms), how the chemical is introduced to the organism, use of environmentally relevant conditions, and standardization of units and methods of normalization to calculate BAF and BSAF values.

CHAPTER 8.0

TOXICITY OF BIOSOLIDS-BORNE TRACE ORGANIC CHEMICALS IN SOILS

8.1 Introduction

8.1.1 Background

Concerns about the toxicity of biosolids-borne trace organic chemicals (TOrCs) in soils arise from land application of biosolids and potential for subsequent exposure of humans and organisms in the environment. As discussed in Chapter 3.0, various potential exposure scenarios can be envisioned:

- Soil - Direct contact (ingestion, dermal) may occur with contaminated biosolids and soils amended with biosolids.
- Water - Contaminants from treated soil may leach into groundwater and surface water, resulting in waterborne exposures (dermal contact, ingestion of drinking water, bioconcentration).
- Sediment - Benthic organisms might be exposed when waterborne contaminants, arising from biosolids partitioning into sediment or binding to particulates, settle into sediments or when contaminated soil washes into surface waters.
- Air - Volatile contaminants occurring in biosolids might vaporize into the atmosphere, where terrestrial organisms might be exposed by inhalation, dermally, or by subsequent wet or dry deposition onto surfaces (e.g., plants exposed by deposition onto leaf surfaces, ingestion of surface-contaminated plants by herbivores, etc.). Wet or dry deposition might also occur on surface waters and soils.
- Biota - Contaminants introduced via biosolids may enter food webs through bioaccumulation and subsequent transfer to higher trophic levels.

8.1.2 Literature Search Strategy

Searches for human toxicity information were limited to identifying reference doses (RfDs), acceptable daily intakes (ADIs), and other human health benchmarks. An ADI is defined as the amount of a chemical to which a person can be exposed on a daily basis over an extended period of time (usually a lifetime) without deleterious effects (U.S. EPA, 2010). The RfD is a similar term, defined as "an estimate (with uncertainty spanning perhaps an order of magnitude) of a daily oral exposure to the human population (including sensitive subgroups) that is likely to be without an appreciable risk of deleterious effects during a lifetime)"(U.S. EPA, 2010). RfDs and ADIs are expressed as intake on a body weight basis (e.g., µg/kg/day or mg/kg/day). U.S.

EPA uses standard assumptions (e.g., 70 kg body weight for an adult) to estimate a RFD, but this may vary depending on the group or agency that develops the toxicity value and the most sensitive subpopulation to be protected (e.g., adults versus infants). Searches for ecotoxicology information focused on obtaining soil toxicity information (top priority, most relevant) and sediment toxicity information (second priority). Sediment toxicity tests were considered only when toxicant doses were delivered by spiking the sediment with test compounds and toxicity data were reported in terms of concentration in sediment. The data sources described in Table 8-1 were identified as those most likely to yield relevant information quickly and thus were searched first.

Table 8-1. Selected Sources of Relevant Toxicity Data.

Data Source	Searched	Human Health	Sediment or Soil Ecotoxicity
ATSDR Minimal Risk Levels (ATSDR, 2009a, citing Tillman 2004)	Yes	X	
ATSDR Toxicological Profiles (ATSDR, 2009b)	Yes	X	X[1]
Cal/Ecotox database (Cal/Ecotox, 2009)	Yes		X
California EPA Public Health Goals	Yes	X	
Contaminant Hazard Reviews (Eisler, 2000a)	Yes	X	X
ECOTOX database (U.S. EPA, 2009c)	Yes		X
Handbook of Chemical Risk Assessment (Eisler, 2000b)	Yes	X	X
Hazardous Substances Data Bank (NIH, 2009)	Yes	X	X
IPCS CICADs (WHO, 2009)	Yes		
Integrated Risk Information System (U.S. EPA, 2009d)	Yes	X	
JECFA acceptable daily intakes (WHO, 2000; ILSI, 2006)	Yes	X	
Risk Assessment Information System (DOE, 2009)	Yes	X	X
SETAC journal *Environmental Toxicology & Chemistry* (SETAC, 2009)	Yes		X
ISI Web of Knowledge	Yes	X[2]	X
UKWIR/WRc toxicity datasheets (UKWIR, 2009)	Yes	X	X
U.S. EPA Drinking Water Standards and Health Advisories (U.S. EPA, 2009b)	Yes	X	
USGS health-based screening levels (USGS, 2009)	Yes	X	

1 In general, only limited ecotoxicology information is available in these reports.
2 This database was searched only for relevant ecotoxicology data, not human toxicology data.

8.2 Overview of Literature Search Results

The following data sources, identified in Table 8-1, contained no information about the high priority contaminants identified for this project: Risk Assessment Information System, Cal/Ecotox database, Contaminant Hazard Reviews, Handbook of Chemical Risk Assessment, the ECOTOX database, and California Public Health Goals. The U.S. EPA Drinking Water Standards and Health Advisories contained relevant information only for perfluorooctanoic acid (PFOA) and perfluorooctane sulfonate (PFOS).

Table 8-2 indicates whether relevant human toxicity or soil or sediment ecotoxicology data were found in the literature for each high priority contaminant. Results of searches for toxicity information for the various classes of selected compounds are described in greater detail below.

Table 8-2. Availability of Relevant Toxicology Data for High Priority TOrCs.

Compound Name	Human Toxicity	Ecotoxicity
Pharmaceuticals and Personal Care Products (PPCPs)		
4-Epitetracycline		
Cimetidine	X	
Ciprofloxacin (CIP)	X	X
Doxycycline	X	
Galaxolide (HHCB)		X
Miconazole		
Ofloxacin		
Tetracycline (TC)	X	X
Tonalide (AHTN)		
Triclocarban (TCC)		
Steroidal Chemicals		
Triclosan (TCS)	X*	X
Ethinyl estradiol (EE2)	X	X
Brominated Fire Retardants		
Mestranol	X	
BDE 28 (2,4,4'-Tribromodiphenyl Ether)		
BDE 47 (2,2',4,4'-Tetrabromodiphenyl Ether)	X	
BDE 85 (2,2',3,4,4'-Pentabromodiphenyl Ether)		
BDE 99 (2,2',4,4',5-Pentabromodiphenyl Ether)	X*	
BDE 100 (2,2',4,4',6-Pentabromodiphenyl Ether)		
BDE 138(2,2',3,4,4',5'-Hexabromodiphenyl Ether)		
BDE 153 (2,2',4,4',5,5'-Hexabromodiphenyl Ether)	X*	
BDE 183 (2,2',3,4,4',5',6-Heptabromodiphenyl Ether)		
BDE 209 (2,2',3,3',4,4',5,5',6,6'-Decabromodiphenyl Ether)	X	X
Dimethyl Tetrabromobisphenol A		
Hexabromocyclododecane (isomers) (HBCD)		X
Tetrabromobisphenol A (TBBPA)		X
Perfluorinated Chemicals (PFCs) and PFC Precursors		
Perfluorooctanoic acid (PFOA)	X	
Perfluorononanoic acid (PFNA)		X
Perfluorodecanoic acid (PFDA)		
Perfluoroundecanoic acid (PFUnDA)		
Perfluorododecanoic acid (PFDoDA)		
Perfluorotetradecanoic acid (PFTeDA)		
FOSA (Perfluorooctane sulfonamide)		
N-EtFOSAA (2-(N-ethylperfluorooctanesulfonamido)acetate)		
N-MeFOSAA (2-(N-methylperfluorooctanesulfonamido)acetate)		
Perfluorohexane sulfonate (PFHxS)		
Perfluorooctane sulfonate (PFOS)	X	
Perfluorodecane sulfonate (PFDS)		
Plasticizers and Surfactants		
Bisphenol A (BPA)	X	X
4-Cumylphenol	X	
4-tert-Octyl phenol	X	X

*Human health benchmark (HHB) available, according to the Memorandum entitled "EPA Contract 68-C-09-001, Task 4, Work Assignment #B-20: Feasibility of modeling analytes identified in the TNSSS."

8.2.1 Ecotoxicity of Selected TOrCs in Soil or Sediment

In general, databases and other data compilations (Table 8-1) were not particularly useful to identify relevant soil or sediment data for the high priority TOrCs identified for this project. These data sources tend to focus on priority pollutants that were not the targets of the current effort.

8.2.2 Human Toxicity of Selected TOrCs

The literature search has uncovered relevant human health toxicity data for several of the targeted TOrCs for this project. Additional information might be found if a more extensive literature review is conducted.

8.3 Literature Search Results for Individual Chemicals or Classes

8.3.1 Pharmaceuticals and Personal Care Products

Extensive reviews of the human and ecological effects of pharmaceuticals and personal care products (PPCPs) have been completed (Snyder et al., 2008a; Snyder et al., 2008b) and are briefly summarized here. PPCPs exhibit a wide range of modes of action, some of which are intended (e.g., therapeutic effects of pharmaceuticals). However, theraepeutic effects are not desired for non-target populations of humans and animals for which pharmaceuticals were not intended. Furthermore, pharmaceuticals have side effects that are not desired even in the populations for which they are intended. As pharmaceuticals undergo testing during the registration process, the toxicological database for these chemicals is typically much more robust than for personal care products. However, toxicity data for nontarget groups are often lacking, and potential for unintended effects or alternative modes of action (e.g., endocrine disruption) may be poorly investigated. Ecotoxicology data are generally sparse for PPCPs. In some cases, modes of action occur in ecological receptors that would not be predicted based on effects in humans and in common laboratory test animals. The list of target compounds for this review is provided below.

Cimetidine

Cimetidine is a drug used in human and veterinary medicine. Summaries of the therapeutic effects, therapeutic doses, and side effects of this drug are available at www.drugs.com and www.medi-vet.com and are summarized as follows. Cimetidine is used in humans to treat and prevent ulcers of the stomach and small intestine and in treating gastroesphageal reflux disease. The chemical may also be used to treat esophagatis caused by gastric reflux and certain conditions that cause increased acid secretion. Cimetidine has similar uses in animals (e.g., horses) and also has been used investigationally as an immunomodulating agent in dogs. Cimetidine is an H2 (histamine) blocker and reduces acidity in the stomach by blocking histamine, a stimulant of the release of acid into the stomach. Cimetidine may increase susceptibility to viruses that cause pneumonia, particularly for those with compromised immune systems. Cimetidine is an United States Food and Drug Adminstration (FDA) Pregnancy Category B drug, which means that is not expected to be harmful to an unborn baby, i.e., animal reproduction studies have failed to demonstrate a risk to the fetus, but there are no adequate and well-controlled studies in pregnant women. Acceptable daily intakes ranging from 5.7 to 19 μg/kg/day were found in the literature.

A Predicted No-effect Concentration (PNEC) of 740 μg/L for cimetidine in water was developed for an aquatic risk assessment (Ayscough et al., 2000). The PNEC value was derived using an assessment factor of 1,000 and relevant acute data, but the original data and the source were not provided. This would suggest an acute toxicity value of 0.740 mg/L, which indicates high acute aquatic toxicity due to waterborne exposure to cimetidine. A later review found no terrestrial or aquatic ecotoxicology data for cimetidine (Jones et al., 2002). No relevant soil or sediment toxicity data were identified in the current review.

Ciprofloxacin and Ofloxacin

Ciprofloxacin (CIP) and ofloxacin are fluoroquinolone antibiotics used in human and veterinary medicine. The drugs are used to treat bacterial infections that cause bronchitis, pneumonia, certain sexually transmitted diseases, urinary tract infections, and prostate infections and may also have off-label uses. Summaries of the therapeutic effects, therapeutic doses, and side effects of the drugs are available elsewhere (www.drugs.com) and are summarized as follows. CIP and ofloxacin are antibacterial drugs that act by inhibiting deoxyribonulcelic acid (DNA) gyrase, thus halting DNA and protein synthesis and killing bacteria. Animal and human cells are less sensitive than bacterial cells to these drugs. Both CIP and ofloxacin are FDA Pregnancy Category C drugs, i.e., animal reproduction studies have shown an adverse effect on the fetus and there are no adequate and well-controlled studies in humans, but potential benefits may warrant use of the drug in pregnant women despite potential risks. Mammalian toxicity of antibiotics is generally low and ADIs for CIP range from 1.6 - 7.1 μg/kg/day. No relevant human toxicity data were identified for ofloxacin.

CIP is reported to have high acute toxicity to the soil bacterium *Pseudomonas putida* (Ayscough et al., 2000). For waterborne exposures, CIP showed high toxicity to the cyanobacterium *Microcystis aeruginosa* (50% effect concentration (EC_{50}) in the range of 5 - 60 μg/L), lesser toxicity to the alga *S. capricornutum*, and low toxicity to fish and daphnids (no effect demonstrated at 100 and 60 mg/L, respectively; Halling-Sorensen et al., 2000). Greater toxicity to prokaryotes than to eukaryotes is expected. A PNEC was estimated at 0.05 μg/L (Halling-Sorensen et al., 2000). A review of pharmaceuticals in the environment (Boxall et al., 2002) identified aquatic toxicity effect concentrations for CIP in the range of 0.005 - 0.08 mg/L and for ofloxacin in the range of 0.01 - 82.8 mg/L, indicating high toxicity for at least some aquatic organisms. No terrestrial toxicity data for either drug were identified, nor were any soil or sediment toxicity data. CIP was implicated as the cause of genotoxicity demonstrated in an *in vitro* bioassay of hospital effluents, but the results might not be predictive of effects *in vivo* (Ayscough et al., 2000).

Doxycycline, Tetracycline, and 4-Epitetracyline

Doxycycline and tetracycline (TC) are tetracycline-type antibiotics (www.drugs.com) used in both veterinary and human medicine. 4-Epitetracycline is a metabolite of TC (Zurhelle et al., 2000). Summaries of the therapeutic effects, therapeutic doses, and side effects of doxycycline and TC are available at www.drugs.com. Tetracycline drugs work by slowing the growth of bacteria in the body. They are used to treat many bacterial infections, rosacea, acne, and some sexually transmitted diseases. Doxycycline is also used in combination with other medications to treat certain amoeba infections. There may be off-label uses for both drugs. Doxycycline and TC are FDA Category D Drugs, which means that "there is positive evidence of human fetal risk based on adverse reaction data from investigational or marketing experience or studies in humans, but potential benefits may warrant use of the drug in pregnant women despite potential risks." Doxycycline and TC should not be used by pregnant women and may pass into breast milk and affect bone and tooth development in a nursing infant. The drugs should not be administered to children less than eight years of age because it can cause permanent discoloration of teeth and affect growth. Allergic reactions may occur following administration of tetracycline drugs.

Residues of TC may be found in products obtained from food animals treated with the drug, and the Joint FAO/WHO Expert Committee on Food Additives (JECFA, 1998) generated an ADI for the drug (Table 8-4). All three ADIs found in the literature for TC were set at the same level, 30 μg/kg/day. ADIs for doxycycline range from 3 - 30 μg/kg/day. No relevant human toxicity data were found for 4-epitetracycline.

TC hydrochloride appears to have low acute toxicity to lake trout (*Salvelinus namaycush*) based on waterborne exposures. Boxall et al. (2002) reported aquatic toxicity effect concentrations for TC in the range of 0.0251-579 mg/L, indicating that it is highly toxic to at least some aquatic organisms (see reference for data on individual species). TC in the low μmol/kg soil range inhibited iron(III) reduction by soil microorganisms (Table 8-3; Thiele-Bruhn and Beck, 2005). No other sediment or soil toxicity data based on concentrations in relevant media were identified for these three drugs, and a previous review (Boxall et al., 2002) found no terrestrial toxicity data for TC. Resistance to TC has been documented in soil bacteria from farmland treated with pig manure slurry, and studies have documented the transport of TC-resistance genes in groundwater under swine production facilities (Boxall et al., 2003).

Galaxolide and Tonalide

Galaxolide (1,3,4,6,7,8-hexahydro-4,6,6,7,8,8-hexamethyl cyclopenta[γ]-2-benzopyran), or hexahydro hexamethylcyclopentabenzopyran (HHCB), and Tonalide (7-acetyl-1,1,3,4,4,6-hexamethyl-1,2,3,4-tetrahydronaphthalene), or acetyl-hexamethyl--tetrahydronaphthalene (AHTN), are polycyclic musks frequently used as fragrances in soaps, perfumes, air fresheners, detergents, fabric softeners, and other household cleaning products (Artola-Garicano et al., 2003). The chemicals are continuously discharged to the sewer systems in many countries, resulting in μg/L concentrations in wastewater treatment plant (WWTP) effluents. HHCB and AHTN are found in air, surface water, seawater, suspended particulate matter, biota, and human adipose tissue and breast milk.

A review conducted by Artola-Garicano et al. (2003) found no data describing the toxicity of HHCB or AHTN to benthic organisms. Risk assessments for HHCB and AHTN (Van de Plassche and Balk, 1997; Balk and Ford, 1999b; Balk and Ford, 1999a) derived PNECs for benthic organisms from aquatic PNECs based on equilibrium partitioning theory because no experimental data were available. The extrapolation assumes that benthic and aquatic organisms are equally sensitive and that uptake is governed by the aqueous phase. Other studies have demonstrated that uptake into oligochaete worms and midge larvae for chemicals with log octanol-water partitioning coefficient (K_{ow}) values similar to HHCB and AHTN occurs mainly from the aqueous phase. The PNEC for HHCB was estimated at 6.8 μg/L, and the PNEC for AHTN was estimated at 3.5 μg/L. Laboratory studies of the toxicity of HHCB and bioconcentration of this compound from water into midges (*Chironomous riparius*) and oligochaete worms (*Lumbriculus variegatus*) agreed with results of the earlier risk assessments, i.e., the previous risk assessments were sufficiently conservative (Artola-Garicano et al., 2003).

Schreurs et al. (2004) reported that HHCB and AHTN are lipophilic and tend to bioaccumulate in aquatic biota. The compounds were shown to bind to and antagonize the zebrafish estrogen receptor and exert dose-dependent anti-estrogenic (antagonistic) effects in vivo in a transgenic zebrafish bioassay.

Miconazole

Miconazole is a human and veterinary drug used to treat fungal infections by altering the permeability of the fungal cell membrane, leading to cell death. Summaries of the therapeutic effects, therapeutic doses, and side effects of this drug are available elsewhere (www.drugs.com, www.medi-vet.com). In humans, miconazole may be administered topically, vaginally, or, for severe systemic infections, parenterally. The chemical is an FDA Pregnancy Category C drug, which means that animal reproduction studies have shown an adverse effect on the fetus and there are no adequate and well-controlled studies in humans, but potential benefits may warrant use of the drug in pregnant women despite potential risks. Use of the vaginal formulation should be avoided during the first trimester of pregnancy. The safety and efficacy of miconazole in children less than 1 year of age has not been sufficiently studied. No relevant toxicity values were found in the literature for miconazole.

Imidazole fungicidal drugs, including miconazole, inhibit steroidogenic cytochromes such as cytochrome P450 c17αhydroxylase, 17,20-hydroxylase (CYP17; Villeneuve et al., 2007) and, thus may exert endocrine disruptive effects in animals (Walsh et al., 2000). No relevant soil or sediment toxicity data were identified in the literature for miconazole.

Triclocarban

Triclocarban (3,4,4-trichlorocarbanilide; TCC) is a bacteriostatic agent widely used in soaps, detergents, and plastics, and is classified as a High Production Volume chemical in the U.S. Toxicology data are sparse. The chemical appears to have low acute oral toxicity in rodents, but high acute dermal toxicity in rabbits. Toxicity data derived from waterborne exposures to triclosan (TCS), suggest high acute and chronic toxicity to both freshwater and marine organisms. In freshwater tests using waterborne exposures, TCC appears to have high acute and chronic toxicity to crustaceans and high acute toxicity to insects and fish. In marine tests using waterborne exposures, the data suggest high acute and chronic toxicity to mollusks and crustaceans. The current review identified no relevant soil or sediment toxicity data and no environmental standards for TCC.

Triclosan

Triclosan (2,4,4'-trichloro-2'-hydroxydiphenyl ether; TCS) is a broad-spectrum antimicrobial used widely around the world in consumer products such as soaps, detergents, surface cleansers, disinfectants, toothpaste and other oral hygiene products, cosmetics and other topical personal care products, and pharmaceuticals. Approximately 96% of the uses of TCS involve consumer products that are discarded through residential drains to municipal sewers. The remaining uses include addition to some plastic formulations and impregnation of surfaces including food wrappers, chopping boards, and refrigerator linings. Because of its primary uses in consumer products, the majority of TCS is disposed in municipal sewer systems and can enter the aquatic environment in effluents and sewage sludges (Orvos et al., 2002; van Wezel and Jager, 2002; Capdevielle et al., 2008; Reiss et al., 2009).

TCS inhibits fatty acid synthesis at the enoyl-acyl carrier protein reductase step in bacterial systems (Snyder et al., 2008a; Snyder et al., 2008b). Based on an ecotoxicological assessment, TCS is classified on the basis of its mode of action as a reactive chemical or a nonpolar narcotic chemical (van Wezel and Jager, 2002). Biological membranes are generally permeable to unionized molecules and relatively impermeable to ionized species of molecules. Consistent with this general rule, the ionized form of TCS is less toxic than the neutral form to algae and daphnids (Capdevielle et al., 2008).

A recent review of toxicological information was conducted for TCS to develop an ADI for humans (Snyder et al., 2008a). No effects are reported in most animal studies, and animal studies have shown no evidence of carcinogenicity. Two studies in rats report non-neoplastic changes in the liver. No adequate and well-controlled studies of TCS exposure have been conducted in pregnant women. The current review identified three relevant toxicity values for TCS, all at 30 μg/kg/day (Table 8-4).

Reiss et al. (2009) conducted an ecological risk assessment for TCS in the terrestrial environment. The assessment addressed the following: direct exposure via biosolids-amended soils to terrestrial plants, earthworms, and soil microorganisms; secondary exposure to birds and mammals from consumption of earthworms exposed to TCS in soil; and secondary exposure to birds and mammals from consumption of fish exposed to TCS in rivers or streams that receive WWTP effluent. The assessment identified no significant risks based on currently available toxicity and occurrence information for TCS. However, the authors noted that only a limited number of plant species was tested, so effects to other plants could not be ruled out at the greatest concentrations reported in biosolids-amended soils (Kinney et al., 2008).

Ecotoxicology data reviewed by Reiss et al. (2009) can be summarized as follows:

- Bacterial toxicity: As a bactericide, TCS has numerous intracellular and cytoplasmic target sites and may influence signaling pathways and transcription of genes involved in amino acid, carbohydrate, and lipid metabolism. The major target is lipid metabolism, and one molecular mechanism by which this occurs involves inhibition of type II bacterial fatty acid synthesis. Type II fatty acid synthesis is important in bacteria and plants.
- Avian toxicity: Based on the results of three acute toxicity studies, TCS exhibits low toxicity to birds. Chronic toxicity data were not available.
- Mammalian toxicity: Acute oral 50% lethal dose (LD_{50}) values range from 3,700 - 5,000 mg/kg body weight, indicating that TCS is practically nontoxic to mammals on an acute basis. There have been numerous animal studies with several species using an oral route of administration, which is most relevant for mammalian risk assessment. The authors focused on the two most recent studies conducted according to Organization for Economic Cooperation and Development (OECD) guidelines. A subchronic oral toxicity study in rats resulted in a no observed adverse effect level of 1,000 ppm (~100 mg/kg body weight/day). A lifetime exposure study with hamsters yielded a no observed effect level (NOEL) of 75 mg/kg body weight/day.
- Toxicity to soil-dwelling organisms: Acute toxicity of TCS was assessed in the composting earthworm (Eisenia foetida foetida), resulting in a 14-day 50% lethal concentration (LC_{50}) > 1,026 mg/kg soil (dry weight) and a no observed effect concentration (NOEC) that was at least the same level. A radiolabeled bacterial toxicity test was conducted on mixed microbial populations (activated sludge mixed liquor) that might be found in WWTPs, septic tanks, or the surface layer of soils. Heterotrophic activity was inhibited at 239 mg/L TCS. Although the bacterial test has been quoted by others as being relevant to the top layers of soil, some suggest it is not relevant to soil-dwelling organisms (Samsøe-Petersen et al., 2003).
- Terrestrial plant toxicity: Two relevant studies were available. One assessed the effect of TCS on seedling emergence and growth (pre-emergent study). The other assessed effects on vegetative vigor. In the pre-emergent study, cucumber seeds were planted in a treated sandy loam soil. NOECs for emergence, shoot length, shoot dry weight, and root dry weight were at least the greatest concentration tested (1mg/kg, nominal value). The vehicle blank affected shoot length and shoot dry weight, lending some uncertainty to the results. The vegetative vigor study involved six plant species, three monocotyledons and three dicotyledons. Cucumber was the most sensitive species, with an EC_{50} of 0.74 mg/kg

TCS in soil for reduction in shoot dry weight. The utility of this study was limited because it was conducted in quartz sand, which is not representative of soils prevalent in agricultural conditions. In comparison, the seedling emergence study conducted in sandy loam soil demonstrated no effects.

- Soil process effects: The effects of TCS on soil nitrification and soil respiration were tested in moist sandy loam soil according to standard OECD guidelines. TCS had no detrimental effect on soil nitrification at concentrations up to 2 mg/kg soil (dry weight) after 28 days of incubation and no detrimental effect on soil respiration (CO_2 generation) after 28 days of incubation up to the greatest concentration tested (2 mg/kg_{dw} soil).

Orvos et al. (2002) reviewed available information regarding the aquatic toxicity of TCS and conducted a number of additional aquatic toxicity tests. TCS was much less toxic to WWTP microbes than to the other types of organisms studied. Narcosis appears to be the mode of action in fish. Algae, cyanobacteria, and other prokaryotes are the most sensitive aquatic organisms to the toxic effects of TCS. Waterborne TCS is acutely toxic to fish (LC_{50}: 0.25-2 mg/L) and daphnids (0.39 mg/L; van Wezel and Jager, 2002). Waterborne TCS is reported to be weakly androgenic in fish (Foran et al., 2000).

Overall, TCS exhibits little toxicity to terrestrial organisms and substantially greater toxicity to aquatic organisms. Terrestrial plants and aquatic algae were the most sensitive species to the effects of TCS. The current review identified a substantial body of relevant soil toxicity data for TCS effects on plants, invertebrates, and soil microorganisms (Table 8-3).

8.3.2 Steroidal Chemicals

17α-Ethinyl estradiol

Snyder et al. (2008a) reviewed the human toxicology of ethinyl estradiol (EE2). More details are available in that report and are summarized here. EE2 is a synthetic pharmaceutical estrogen used most frequently as the estrogenic component in combined oral contraceptives. As such, is known to disrupt the endocrine system and was designed for that purpose. It is also used to treat menopausal and post-menopausal symptoms, female hypogonadism, and malignant neoplasm of the breast and prostate, acne, and Turner's syndrome. EE2 appears to act primarily by binding to the estrogen receptor and to exert its main therapeutic and toxic or adverse effects through that mechanism. Therapeutic estrogen use has side effects, including increases in incidence of gall bladder disease and induced adverse effects on coagulation. In 1987, the International Agency for Research on Cancer found sufficient evidence for carcinogenicity of steroidal estrogens, including EE2, in humans. Endometrial, breast, and certain liver cancers may occur at higher incidence in people treated with EE2 than in the general population. In 1988, the State of California listed EE2 among those chemicals known to the state to cause cancer or *reproductive toxicity.*

No minimal lethal or toxic dose of EE2 has been established, and there is no dose of EE2 that is both therapeutically effective and free of side effects. For this reason, minimally effective doses are recommended for long-term use. The main risks and target organs for chronic toxicity of EE2 are hypertension, fluid retention, cardiovascular disease, cerebrovascular disease, thromoembolic disease, gallbladder disease, and certain cancers in some people. Fetuses and prepubertal children are considered to be potentially sensitive subpopulations. EE2 is an FDA Pregnancy Category X drug, meaning that its use is contraindicated during pregnancy. It should not be used during pregnancy due to the potential for mutagenicity and teratogenicity. However, to date, studies of children accidentally exposed in utero to EE2 have shown no adverse effects.

EE2 is present in breast milk of women who take oral contraceptives, and there have been cased where jaundice and breast enlargement were reported in infants of women who nursed them while taking oral contraceptives. A study of children who were nursed by mothers who used oral contraceptives during lactation revealed no effects on intellectual and psychomotor behavior or height and weight increases. Estrogens should be used in children with caution because estrogens can enhance epiphyseal plate closure in bone and affect growth. EE2 can increase levels of corticosteroid-binding globulin in post-menopausal women, and authors of a study on this subject appeared to consider this an adverse effect. EE2 reportedly suppresses follicle-stimulating hormone in post-menopausal women.

Animal studies suggest that EE2 exposure can cause endocrine disruption and alterations in sexual development in rodents exposed *in utero*. For example, investigators in one study concluded their results demonstrate that developmental exposure to oral micromolar doses of EE2 can permanently disrupt the reproductive tract of the male rat (Howdeshell et al., 2008). The subject of the potential for developmental exposure to very small doses of EE2 to affect reproductive endpoints such as androgen-dependent tissue (e.g., prostate) weights has been highly controversial, in part due to the inability of some laboratories to replicate results indicating low dose effects. The controversy prompted the National Toxicology Program (2001) to convene an expert panel charged with evaluating studies that might suggest that very low doses of EE2 can cause endocrine disruptive effects in rodents exposed *in utero*.

The Australian Environmental Protection and Heritage Council (EPHC, 2008) used the lowest therapeutic dose to derive surrogate TDIs (s-TDIs) for synthetic and natural hormones without previously established ADIs, including EE2. The s-TDI for EE2 is 4.3×10^{-5} μg/kg/day. The value was based on the therapeutic dose, which is not without side effects, but safety factors were applied in the development of the s-TDI. Also, estrogen use is contraindicated in some groups of people, so its effects are not desired in these populations. Snyder et al. (2008a) derived an ADI of 0.00010 μg/kg/day for EE2.

There is a substantial body of literature investigating the aquatic toxicity of EE2. Notably, a study was conducted to investigate the effects of EE2 on a population of fish exposed in an experimental lake over seven years (Kidd et al., 2007). The investigators concluded that "chronic exposure of fathead minnow (*Pimephales promelas*) to low concentrations (5-6 ng/L) of the potent 17α-ethynylestradiol led to feminization of males through the production of vitellogenin mRNA and protein, impacts on gonadal development as evidenced by intersex in males and altered oogenesis in females, and, ultimately, a near extinction of this species from the lake." However, these concentrations are not environmentally relevant. Environmentally relevant concentrations of EE2 (less than 1 ng/L) in water have been reported to reduce egg fertilization success and cause demasculination in fathead minnows (Parrott and Blunt, 2005)

Mestranol

Mestranol is a prodrug that is metabolized to an active form, ethinyl estradiol, in the body. Mestranol is an estrogenic drug, and, as discussed for EE2, the International Agency for Research on Cancer found sufficient evidence to indicate that post-menopausal estrogen therapy is carcinogenic in women. The Australian Environmental Protection and Heritage Council used the lowest therapeutic dose to derive surrogate TDIs (s-TDIs) for synthetic and natural hormones without previously established ADIs, including mestranol (EPHC, 2008). The s-TDI for mestranol is 7.1×10^{-5} μg/kg/day. The value was based on the therapeutic dose, which is not without side effects, but safety factors were applied in the development of the s-TDI.

8.3.3 Brominated Fire Retardants

Brominated fire retardants (BFRs) are substances used in plastics, textiles, electronic circuitry, and other materials to prevent fires. Thorough reviews of the human and ecological toxicity of BFRs are available elsewhere (de Wit, 2002).

Polybrominated Diphenyl Ethers

Polybrominated diphenyl ethers (PBDEs) are a class of fire retardants added to products commonly found in homes, offices, automobiles, and airplanes. There are theoretically 209 possible PBDE congeners. Historically, three technical brominated diphenyl ether (BDE) mixtures were primarily produced: penta BDE, octa BDE, and deca BDE. Only deca BDE is still produced and used at high volumes, but most products containing PBDEs are used for years and often for more than a decade, releasing PBDEs into the environment during their lifetime. PBDEs migrate from these products into the environment and are now ubiquitous contaminants found in indoor and outdoor air, dust in homes and offices, food, surface water, remote Arctic regions, terrestrial and marine mammals, fish, and nearly all people examined in biomonitoring studies (McDonald, 2005).

BDE 47 and BDE 99 are the most frequently encountered PBDEs. The chemicals have been found in environmental samples (e.g., water and sediment), in various wildlife and aquatic species (e.g., pike, trout, and salmon), and in human adipose tissue, blood, and breast milk (Norén and Meironyté, 2000; Evandri et al., 2003). PBDEs are lipophilic, and infant exposure is suspected to occur by placental transfer and breast milk ingestion (Vonderheide et al., 2008). Primary sources of human exposure to PBDEs include dust inhalation, occupational exposure, and dietary intake, particularly of foods rich in fat (meat, dairy products, and especially fish; Vonderheide et al., 2008). While there are many reports of PBDEs in air and sediments, relatively little is known about PBDEs in surface waters and soils (Vonderheide et al., 2008). Fate models predict that soil, the least studied environmental compartment with regard to PBDE contamination, is the most likely environmental sink (Vonderheide et al., 2008). Penta BDE congeners tend to be dominant in the atmosphere and in water, while BDE 209 is more important in soil, sediment, and sewage sludge (Vonderheide et al., 2008). Five congeners (BDE 47, BDE 99, BDE 100, BDE 153, and BDE 154) predominate in human tissues, usually accounting for more than 90% of the total PBDE body burden in most individuals who are not exposed occupationally (McDonald, 2005). Like polychlorinated biphenyls (PCBs) and dioxins, the lower-brominated PBDEs have long half-lives (2-12 years, depending on the congener) in the human body, while half-lives in rodents are much shorter.

Penta-, octa-, and deca BDE technical mixtures are not acutely toxic (McDonald, 2005). PBDE technical products have thyroid disrupting properties (Vonderheide et al., 2008). All PBDE technical mixtures bind competitively to the thyroid hormone receptors, probably due to structural similarities with thyroid hormones, and PBDEs can alter liver function, producing changes in vitamin A homeostasis and thyroid hormone levels (often increased elimination of thyroxine; Vonderheide et al., 2008). Long-term administration to rodents resulted in liver enzyme induction and thyroid hormone disruption, with deca-BDE less effective than penta- or octa BDE (McDonald, 2005). Penta BDEs can interfere with sexual development (delayed onset of puberty, decreased follicle formation) and sexually dimorphic behaviors (increased sweet preference in males; Vonderheide et al., 2008). Penta BDEs also bind to the androgen receptor. Competitive binding to the androgen receptor and anti-androgenic properties of penta BDEs has been implicated in delayed onset of puberty and decrease in the size of androgen-dependent tissues in exposed animals (Vonderheide et al., 2008). Preliminary research suggests that perinatal exposure to PBDEs might alter later cognitive function in the same manner as PCBs (Vonderheide et al., 2008). PBDEs have also been shown to exert cytotoxic effects (Vonderheide et al., 2008). Concern regarding the carcinogenicity of deca BDEs is low. However, based on similarities to other contaminants, PBDE congeners with lesser bromination can be expected to be carcinogenic, and exposure to these has been linked to tumor formation and cancer (Vonderheide et al., 2008).

Health effects occurring at the least exposures in animal studies appear to be developmental effects, including harm to the brain and reproductive organs (McDonald, 2005). The U.S. EPA identified neurobehavioral effects as the critical effects in the development of reference doses for (deca-) BDE 209, (penta-) BDE 99, (tetra-) BDE 47, and (hexa-) BDE 153, with RfDs in the low μg/kg/day range (see Table 8-4). No relevant human toxicity values for other individual PBDEs or for hexabromocyclododecane (HBCD) or dimethyltetrabromobisphenol A were identified.

According to Norén and Meironyté (2000), the exponential increase of PBDEs in human breast milk poses a risk to infants and young children. To characterize the potential health risks posed by PBDEs, McDonald (2005) compared tissue concentrations of the sum of five prominent congeners found in people to tissue concentrations in rodents resulting from the highest doses that caused no developmental toxicity. He concluded that if humans are as sensitive as animals to developmental toxicity induced by PBDEs, the margin of safety is low for a fraction of the human population.

A recent review of the human toxicity of PBDEs (JECFA, 2005) concluded that PBDEs are non-genotoxic substances. The review committee declined to propose a provisional tolerable daily intake level because 1) PBDEs are comprised of a complex group of related chemicals, and a single technical mixture or consistent pattern of congeners could not be identified in food; 2) there were inadequate data to establish a common mechanism of action for PBDE congeners so that one congener could be used as a surrogate for total exposure; 3) the database on toxicity of the main congeners found in the diet was insufficient to define NOELs; 4) the toxicological signficance of some of the reported effects was unknown; and 5) toxicity testing on purified individual congeners in vitro indicated no aryl hydrocarbon receptor (Ah) receptor activation, but many of the reported adverse effects are similar to those exerted by dioxin-like contaminants, suggesting that highly potent, trace dioxin-like impurities in some PBDE test compounds or mixtures may be responsible for the effects. However, the committee also stated that despite that paucity of data on toxicity and intake levels, there appeared to be a large margin of exposure for for a non-genotoxic compound, indicating that intake of PBDEs is not likely to pose a health risk. More information about the toxicity of individual PBDE congeners can be found in the review by de Wit (2002) and references therein.

The current review identified only one publication containing relevant soil or sediment toxicity data for PBDEs. Sverdrup et al. (2006) investigated the toxicity of decabromodiphenyl ether to a soil invertebrate, red clover, and soil nitrifying bacteria (Table 8-3). The test chemical was not toxic to any of these organisms at concentrations of at least 1000 mg/kg soil (dry weight).

Tetrabromobisphenol A

Canesi et al. (2005) investigated the effects of tetrabromobisphenol A (TBBPA) in mussels and characterized the toxicity of the compound as follows. TBBPA is the most important individual BFR. Although the compound and its derivatives have been found in environmental samples, data are very limited on the presence of this compound in biota. Research on mammals indicates that it has low toxicity, but *in vitro* studies demonstrate activity as a cytotoxicant, neurotoxicant, immunotoxicant, and thyroid hormone agonist, and indicate weak estrogenic activity. The effects of TBBPA have been recently ascribed to its interactions with cellular signaling pathways, in particular with mitogen activated protein kinases (MAPKs). The compound has high acute toxicity to aquatic organisms, such as algae, mollusks, crustaceans, and fish, but little is known about the mechanisms of action of this compound aquatic species. A study of the effects of TBBPA on immune cells (hemocytes) of the marine mussel *Mytilus galloprovincialis* demonstrated that it induces hemocyte lysosomal membrane destabilizatoin in the low micromolar range, and the effect was reduced or prevented by hemocyte pre-treatment by specific inhibitors of MAPKs and of protein kinase C. According to the investigators, TBBPA stimulated phosphorylation of MAPK members and protein kinase C and significantly stimulated the hemocyte microbiocidal activity towards *E. coli*, lysosomal enzyme release, phagocytic activity, and extracellular superoxide production. The results demonstrate that TBBPA activates the immune function of mussel hemocytes *in vitro* through kinase-mediated cell signaling and that common transduction pathways are involved in mediating the effects of this BFR in mammalian and aquatic invertebrate cells.

The current literature search identified no relevant human toxicology data. The search for soil and sediment ecotoxicology data revealed only one publication containing relevant data (Table 8-3). TBBPA appears to be relatively nontoxic to red clover and soil nitrifying bacteria and toxic to a soil invertebrate. The authors predicted a no-effect concentration of 0.3 mg/kg soil (dry weight).

8.3.4 Perfluorochemicals and PFC Precursors

Hekster et al. (2003) reviewed of the state of knowledge regarding the environmental toxicology of perfluoroalkylated substances. The following discussion is taken mainly from that review and from information available at www.drugs.com.

Reports on the occurrence of perfluorochemicals (PFCs) in the environment and in occupational settings have drawn increased scientific and regulatory scrutiny of these chemicals since the 1990s. Most studies of perfluorinated compounds have focused on PFOS and PFOA.

The mechanisms of toxicity for individual PFCs are poorly understood. The perfluorocarboxylates (PFCAs), including PFOA, are peroxisome proliferators (Hekster et al., 2003). Peroxisome proliferators are chemicals that interact with the peroxisome proliferator-actived receptors (PPARs; Vanden Heuvel, 2009). PPAR isotypes have been identified in mammals and lower vertebrates. PPARs are nuclear receptors that act as ligand-activated transcription factors controlling many cellular and metabolic processes such as energy homeostasis and inflammatory responses (Vanden Heuvel, 2009). Their activity can be modulated by drugs such as hypolipidemic fibrates (e.g., clofibrate) and insulin sensitizing thiazolidinediones (e.g., rosiglitazone; tradename: Avandia).

The current search identified no soil or sediment toxicology data for the perfluorinated compounds except for perfluorononanoic acid (PFNA). PFNA was relatively nontoxic to a sediment-dwelling polychaete (Table 8-3).

Perfluorooctanoic Acid

PFOA is distributed to the liver, plasma, and kidney and is excreted slowly from the body via urine and feces. PFOA is not acutely toxic to rodents. In chronic feeding studies with rodents, the primary target organ for toxicity of PFOA was the liver. As discussed previously, PFOA is a peroxisome proliferator (Hekster et al., 2003).

The U.S. EPA set a subchronic RfD for PFOA at 0.0002 mg/kg/day based on liver effects (U.S. EPA, 2009e). The Minnesota Department of Health set a chronic non-cancer RfD at 0.000077 mg/kg/day (Table 8-4). Using a previously developed, less stringent reference dose (0.001 mg/kg/day), the Minnesota Department of Health (2008; 2009b; 2009a) developed a Residential Soil Reference Value of 30 mg/kg and an Industrial Soil Reference Value of 200 mg/kg. Soil concentrations less than these amounts should provide an adequate level of protection from exposure due to direct contact with contaminated soil. The values are considered to be provisional and do not address impacts to groundwater, food chain impacts, or ecological effects.

Perfluorooctane Sulfonate

Like PFOA, PFOS is distributed to the liver, plasma, and kidney and is excreted slowly from the body via urine and feces. PFOS is not acutely toxic to rodents. The U.S. EPA set a subchronic reference dose for PFOS at 0.00008 mg/kg/day based on thyroid system disruption and reduced levels of high-density lipoproteins (U.S. EPA, 2009e). Likewise, the Minnesota Department of Health (2008; 2009b; 2009a) set a reference dose for PFOS at 0.00008 mg/kg/day based on developmental, hepatic, and thyroid effects. Using a previously developed, less stringent reference dose (0.0002 mg/kg/day), the Minnesota Department of Health (2005) developed a Residential Soil Reference Value of 6 mg/kg and an Industrial Soil Reference Value of 40 mg/kg.

Avian toxicity reference values have been developed for PFOS in dietary, mean serum, liver, and egg concentrations (Newsted et al., 2005). The current literature search uncovered no relevant soil or sediment ecotoxicity values.

8.3.5 Surfactants and Plasticizers

A Several surfactants and plasticizers have been investigated for activity as potential endocrine disrupting chemicals (EDCs), including the target compounds for this review.

Bisphenol A

Snyder et al. (2008a) summarized human toxicity data for bisphenol A (BPA). The following information was summarized from that report.

BPA is an intermediate used in the production of epoxy and polycarbonate resins (GWRC, 2003). Polycarbonate plastics are commonly used in the automotive, construction, packaging, and electronics industries. Epoxy resins are used for surface coatings such as paint, linings for metal food cans, bottle tops, dental coatings, and some linings applied to water mains, as well as in the construction and electronics industries (GWRC, 2003). Widespread exposure of humans to BPA through multiple routes is expected. Exposures to BPA may occur from air, water, dust and food. According an EU Risk Assessment Report, the highest potential for human exposure is through products that directly contact food such as food and beverage containers with internal epoxy resin coatings and through the use of polycarbonate tableware and bottles such as those used to feed infants. Following ingestion, BPA is readily absorbed, and both the parent compound and the metabolite are widely distributed (Snyder et al., 2008a). BPA has been found in fetal fluids and in milk in both humans and experimental animals (NTP-CERHR, 2007).

BPA has been the subject of intense study because it is considered to be an EDC with widespread human exposure, including exposure to sensitive subpopulations like infants exposed to BPA leaching from plastic bottles and toys. Several expert groups list BPA as an EDC (EC-BKH, 2000; WHO-IPCS, 2002; GWRC, 2003; IEH, 2005). BPA reportedly possesses estrogenic activity and affinity for the estrogen receptor *in vitro* (IEH, 2005). EC-BKH (2000) listed BPA among those chemicals associated with at least one study providing evidence of endocrine disruption in an intact organism and for which there is concern for potential exposure and effects in humans.

Based on a 1982 study by the National Toxicology Program, the U.S. EPA set the oral RfD at 0.05 mg/kg/d based on the lowest observed adverse effect level of 50 mg/kg/day for reduced mean body weight in rats. This is the basis for the reference dose in the U.S. EPA's Integrated Risk Information System (IRIS) database. The U.S. EPA's oral RfD assessment was last revised in 1993 (U.S. EPA, 2009a). U.S. EPA stated that the developmental toxicity of BPA was adequately investigated and expressed high confidence in the RfD. BPA has not undergone a complete evaluation and determination under the U.S. EPA's IRIS program for evidence of human carcinogenic potential.

Numerous organizations have reviewed the toxicity of BPA since U.S. EPA set its RfD. Interest was spurred by studies describing its estrogenic activity, concerns over the potential for reproductive and developmental effects, and potential for exposure. An expert panel convened by the NTP Center for the Evaluation of Risks to Human Reproduction evaluated studies of the reproductive and developmental effects of BPA (NTP-CERHR, 2007). The panel findings may be summarized as follows. Data in mice and rats are sufficient to conclude that BPA causes female and male reproductive toxicity with subchronic or chronic oral exposures. It does not cause malformation or birth defects in rats or mice at levels up the highest doses evaluated at that time. BPA causes neural and behavioral alterations related to sexual dimorphism in rats and mice, and although this raised some concern it is not clear whether the reported effects can be considered adverse toxicological responses. Many of the studies indicating that BPA can cause endocrine-mediated effects at very small doses in rodents are controversial because the results have not been replicated in subsequent studies (NTP-CERHR, 2007). Populations that may be more sensitive to BPA are pregnant and lactating women, embryos, fetuses, infants, and children (NTP-CERHR, 2007).

More recently, Snyder et al. (2008a) calculated an ADI of 50 μg/kg-day for BPA based on developmental toxicity studies that evaluated effects in offspring of rodents exposed to BPA *in utero*. These studies found that offspring of dams exposed from gestation days 11-17 had increased prostate weights and reduced seminal vesicle weights at a dose of 0.002 mg/kg-day. Several other organizations have recently reviewed the toxicological effects of BPA and published draft screening levels, tolerable daily intake levels, etc. (Snyder et al., 2008a). The current literature review identified relevant human toxicity data ranging from 10-50 μg/kg/day (Table 8-4).

In a multi-generation sublethal assay using the nematode *Caenorhabditis elegans* exposed in agar plates (Nematode Growth Medium), exposure to BPA significantly decreased fecundity. Occasionally, exposed individuals exhibited an abnormal vulva, suggesting that BPA is a reproductive endocrine disrupter in these worms. Because agar was the exposure medium, this relevance of this test for prediction of effects in soil or sediment is unclear. The current literature review identified a few studies of soil and sediment toxicity for BPA. The soil toxicity tests suggest that BPA is moderately toxic to nontoxic to terrestrial isopods. The sediment toxicity tests indicate that BPA is highly toxic to benthic organisms. The differences between the toxicity assessments for studies in soil versus those in sediment may be due in part to the use of sensitive reproductive endpoints for endocrine disruption in the latter studies.

4-Cumylphenol

4-Cumylphenol (p-cumylphenol) is an alkylphenolic chemical and chemical intermediate widely used as a material for polycarbonate plastics, surfactants, fungicides, and preservatives (Nakazawa et al., 2009). Newborn rats treated with 4-cumylphenol by gavage developed kidney effects and alterations in the histology of the ovary and uterus, probably due to estrogenic activity (Nakazawa et al., 2009). These authors also found that p-cumylphenol induced multiple renal tubular cysts in newborn rats. According to Biggers and Laufer (2004), 4-cumylphenol is a known xenoestrogen (estrogenic EDC).

The current literature search identified one relevant toxicity value for humans, a threshold of toxicological concern of 1.5 μg/kg/day (Table 8-4). No relevant soil or sediment ecotoxicology data were identified.

4-Cumylphenol showed strong juvenile hormone activity in a rapid and sensitive juvenile hormone bioassay based on effects on settlement and metamorphosis of larvae of the polychaete *Capitella* (Biggers and Laufer, 2004).

4-tert-Octylphenol

Snyder et al. (2008a) summarized human toxicity data for 4-tert-octylphenol. The following information was distilled from that report.

Octylphenol exists as several isomers. 4-tert-Octylphenol (CASRN 140-66-9) is the most commercially important isomer. It is used as a chemical intermediate, mainly in the production of phenolic resins and lacquers and also in smaller amounts in the production of octylphenol ethoxylates. Octylphenol ethoxylates are commonly used as surfactant additives in the manufacture of plastics and detergents. Octylphenol occurs in the aquatic environment mainly by introduction through sewage effluents as a result of incomplete degradation of octylphenol ethoxylates. 4-*tert*-Octylphenol was found in drinking water in Germany (Kuch and Ballschmiter, 2001).

Studies of reproductive, developmental, and endocrine disruptive effects are available. 4-tert-Octylphenol is purported by several expert groups to be an estrogenic EDC (EDSP, 2005; IEH, 2005; Snyder et al., 2008a) and is included in the Global Water Research Coalition priority list of EDCs (GWRC, 2003). EC-BKH (2000) listed 4-tert-octylphenol among those chemicals associated with at least one study demonstrating endocrine disruption in an intact organism and for which there is concern for potential exposure and effects in humans. 4-*tert*-Octylphenol has been reported to affect early development of animals in some toxicity studies, so infants and children may be more sensitive to exposure than adults (Snyder et al., 2008a). However, Tyl et al. (1999) conducted a 2-generation study in rats according to U.S. EPA's Office of Pollution Prevention and Toxic Substances (OPPTS) Draft Testing Guidelines. The study was accepted by the U.S. EPA as definitive for showing the lack of reproductive and low dose effects of 4-*tert*-octylphenol. According to Van Miller and Staples (2005), "This study is considered to be the most critical and definitive evaluation of potential impact of estrogen-like activity on the overall hazard assessment for octylphenol in mammals." Snyder et al. (2008a) calculated an ADI of 150 µg/kg-day based on reproductive developmental toxicity in rats. This was the only relevant human toxicity value identified during the current literature review.

The current review uncovered only two relevant sediment toxicity studies for 4-tert-octylphenol (Table 8-3). Both studies indicate that this compound is highly toxic to mollusks, possibly because the tests employed highly sensitive endpoints used to assess disruption of endocrine and reproductive function. No relevant soil toxicity data were identified.

8.4 Summary and Conclusions

The current review identified relevant human toxicity values for less than half of the targeted TOrCs. However, the literature search was necessarily limited in scope due to budget and time constraints. A more comprehensive search might uncover additional data. Furthermore, the data that were gathered should be further scrutinized with regard to the confidence that they should engender. For example, substantial bodies of data and expert scientific review were involved in the development of toxicity values for chemicals like PFOA, PFOS, and BPA, but it is likely that relatively little chemical-specific information was available for development of the 4-cumylphenol threshold of toxicological concern. The human toxicity data should also be described in terms of relative toxicity and likelihood of exposure through application of biosolids to soils. For some of these compounds, the known and potential modes of action should be described in more detail based on available information.

Table 8-3. Summary of Soil and Sediment Ecotoxicology Data for High Priority TOrCs.

Chemical	Test organism	Study Description	Effect	Effect Concentration	Data Source
EE2	Marine netted whelk (*Nassarius reticulatus*)	Sediment exposure, 4 weeks	Mortality	EC_{10}: 2.2 µg/kg_{dw} EC_{50}: 28.9 µg/kg_{dw}	(Duft et al., 2007, citing Tillman 2004)
HBCD	European flounder (*Platichthys flesus*)	Chronic exposure via spiked sediment, 78 days	General health and toxicity parameters, aromatase activity in gonad, plasma thyroid hormone levels, hepatic microsomal enzyme activities, vitellogenin induction	NOEC: >8,000 µg/g total organic carbon in sediment (Note: fish exposed to additional test concentrations were provided test compound in food as well as in sediment and so were excluded here.)	(Kuiper et al., 2007)
HHCB	Infaunal deposit-feeding polychaete worm (*Capitella* sp.)	Life table response test, 120 days	Adult survival, age at first reproduction, length of reproductive period, number of broods, individual worm body volume, body-size specific egestion rate	NOEC ≥ 168 mg/ kg_{dw} (sediment) No effects observed	(Ramskov et al., 2009)
HHCB	Infaunal deposit-feeding polychaete worm (*Capitella* sp.)	Life table response test, 120 days	Juvenile survival, brood size	NOEC: 26 mg/ kg_{dw} (sediment) LOEC: 123 mg/ kg_{dw} (sediment)	(Ramskov et al., 2009)
HHCB	Infaunal deposit-feeding polychaete worm (*Capitella* sp.)	Life table response test, 120 days	Maturation time	NOEC: 123 mg/ kg_{dw} (sediment) LOEC: 168 mg/ kg_{dw} (sediment)	(Ramskov et al., 2009)
HHCB	Infaunal deposit-feeding polychaete worm (*Capitella* sp.)	Life table response test, 120 days	Total number of eggs produced, time between breeding attempts (marginally increased)	NOEC: 1.5 mg/ kg_{dw} (sediment) LOEC: 26 mg/ kg_{dw} (sediment)	(Ramskov et al., 2009)
TC	Soil microbial population	Microbial inhibition test in six different topsoils	Microbial iron(III) reduction	ED_{90}: 21- >2,200 µmol/kg soil[4] ED_{50}: 6.1- 520 µmol/kg soil[4] ED_{10}: 2.0- 9.4 µmol/kg soil[4]	(Thiele-Bruhn and Beck, 2005)
TCS		Microbial activity test in two Australian soils - OECD protocol for measuring substrate-induced nitrification by monitoring transformation of added ammonium to nitrate	Nitrification	NOEC: 1 mg/kg soil (sandy soil) LOEC: 5 mg/kg soil (sandy soil) NOEC: 10 mg/kg soil (clay soil) LOEC: 50 mg/kg (clay soil)	(Waller and Kookana, 2009)
TCS	Compost worm (*Eisenia fetida*)	Acute toxicity test in soil, 14 days Study conditions approximated OECD TG No. 207, with deviations from protocol[5]	Survival, weight	LC_{50}: >1,026 mg/kg_{dw} (soil) (no effect) NOEC: ≥1,026 mg kg_{dw} (soil) (no effect)	(Samsøe-Petersen et al., 2003; Reiss et al., 2009) both apparently citing the same unpublished data[6]

a see footnote descriptions on last page of Table.

Table 8-3. Summary of Soil and Sediment Ecotoxicology Data for High Priority TOrCs (continued).

Chemical	Test organism	Study Description	Effect	Effect Concentration	Data Source
TCS	Cucumber	Acute seedling growth test (pre-emergent) Medium: sandy loam soil Effect concentrations based on initial measured values in soil (Reiss et al., 2009)	Emergence and growth (shoot length, shoot dry weight, root dry weight)	NOEC/NOEL: >424-1,000 µg/kg or 1,000 µg/kg[8] LOEC: >1,000 µg/kg (no effect)	(Samsøe-Petersen et al., 2003; Reiss et al., 2009) both apparently citing the same unpublished data[9]
TCS	Cucumber (*Cucumis sativus* L.)	Seedling growth test	Growth: inhibition of root elongation, shoot growth	NOEC: 10 mg/kg (soil) LOEC: 30 mg/kg (soil) EC_{10}: 6 mg/kg (soil) EC_{50}: 108 mg/kg (soil)	(Liu et al., 2009)
TCS	Rice (*Oryza sativa* L.)	Seedling growth test	Growth: inhibition of root elongation, shoot growth	NOEC: 1 mg/kg (soil) LOEC: 10 mg/kg (soil) EC_{10}: 27 mg/kg (soil) EC_{50}: 57 mg/kg (soil)	(Liu et al., 2009)
TCS	Six plant species: corn, ryegrass, wheat, cucumber, soybean, tomato	Acute seedling growth test Medium: quartz sand Effect concentrations based on initial measured values in soil (Reiss et al., 2009)	Growth: shoot length, shoot weight, root weight	NOEC (for cucumber, the most sensitive species): 96 µg/kg LOEC (for cucumber, the most sensitive species): 280 µg/kg EC_{50}: 736 µg/kg (or 0.74 mg/kg) See Reiss et al. (Reiss et al., 2009) for results for other, less sensitive test species	(Samsøe-Petersen et al., 2003; Reiss et al., 2009) both apparently citing the same unpublished data[7]
TCS	Soil microbial community	Soil respiration test, 22 days	Inhibition of soil respiration	10 mg/kg_{dw} (soil)[10]	(Liu et al., 2009)
TCS	Soil microbial community	Phosphatase activity test, 22 days	Inhibition of phosphatase activity	0.1 - 50 mg/kg_{dw} (soil)[11]	(Liu et al., 2009)
TCS	Soil microbial community	Microbial activity test in two Australian soils - OECD protocol for measuring substrate-induced respiration by monitoring carbon dioxide evolution	Respiration	NOEC: >100 mg/kg soil (sandy soil) NOEC: 5 mg/kg soil (clay soil) LOEC: 10 mg/kg soil (clay soil)	(Waller and Kookana, 2009)
BDE 209[2]	Red clover (*Trifolium pratense*)	Soil toxicity test, 21 days Conducted according to OECD Guidelines 208: Terrestrial plants, Growth test (OECD 1984)	Seed emergency, growth	NOEC: >1,000 mg/kg_{dw} (soil)[3] LOEC: >1,000 mg/kg_{dw} (soil)[3]	(Sverdrup et al., 2006)
BDE 209[2]	Soil invertebrate (*Enchytraeus crypticus*)	Soil toxicity test, 21days ISO standardized procedure	Survival, reproduction	NOEC: >1,000 mg/kg_{dw} (soil)[3] LOEC: >1,000 mg/kg_{dw} (soil)[3]	(Sverdrup et al., 2006)

[a] see footnote descriptions on last page of Table.

Table 8-3. Summary of Soil and Sediment Ecotoxicology Data for High Priority TOrCs (continued).

Chemical	Test organism	Study Description	Effect	Effect Concentration	Data Source
BDE 209[2]	Soil nitrifying bacteria	Soil nitrification test, 4 week	Sublethal toxicity	NOEC: >1,000 mg/kg_{dw} (soil)[3] LOEC: >1,000 mg/kg_{dw} (soil)[3]	(Sverdrup et al., 2006)
TBBPA	NA	NA	PNEC for soil organisms	PNEC: 0.3 mg/kg_{dw} (soil)[1]	(Sverdrup et al., 2006)
TBBPA	Red clover (*Trifolium pratense*)	Soil toxicity test, 21days Conducted according to OECD Guidelines 208: Terrestrial plants, Growth test (OECD 1984)	Seed emergency, growth	NOEC: >1,000 mg/kg_{dw} (soil)[1] LOEC: >1,000 mg/kg_{dw} (soil)[1]	(Sverdrup et al., 2006)
TBBPA	Soil invertebrate, enchytraeid (*Enchytraeus crypticus*)	Soil toxicity test, 21days Conducted using ISO standardized procedure	Survival, reproduction	NOEC: 3 mg/kg_{dw} (soil)[1] LOEC: 10 mg/kg_{dw} (soil)[1] EC_{10}: 2.7 mg/kg_{dw} (soil)[1]	(Sverdrup et al., 2006)
TBBPA	Soil nitrifying bacteria	Soil nitrification test, 4 week	Sublethal toxicity	NOEC: 300 mg/kg_{dw} (soil)[1] LOEC: 1,000 mg/kg_{dw} (soil)[1] EC_{10}: 295 mg//kg_{dw} (soil)[1]	(Sverdrup et al., 2006)
PFNA	Sediment-dwelling polychaete (*Hediste diversicolor*)	Sediment toxicity test	Energy storage parameters, total energy stored	NOEC: >500 pg/g_{dw} (sediment)	(Stomperudhaugen et al., 2009)
PFNA	Sediment-dwelling polychaete (*Hediste diversicolor*)	Sediment toxicity test	Energy consumption	NOEC: 40-200 pg/g_{dw} (sediment) LOEC: 200-500 pg/g_{dw} (sediment) Effects were not marked.	(Stomperudhaugen et al., 2009)
4-tert-Octyl phenol	Freshwater mudsnail (*Potamopyrgus antipodarum*)	Sediment toxicity test, 8 weeks	Embryo production (Mortality and total embryo number were also assessed)	EC_{10} (4 week): 0.004 μg/ kg_{dw} (sediment) EC_{50} (4 week): 0.07 μg/ kg_{dw} (sediment) LOEC: 1 μg/ kg_{dw} (sediment)	(Duft et al., 2007, citing Duft 2003)
4-tert-Octyl phenol	Marine netted whelk (*Nassarius reticulatus*)	Sediment toxicity test, 4 weeks	Increased weight of gland complex (Imposex and mortality were also assessed)	EC_{10}: 4.3 μg/ kg_{dw} (sediment) EC_{50}: 40.7 μg/ kg_{dw} (sediment)	(Duft et al., 2007, citing Tillman 2004)
BPA	Freshwater mudsnail (*Potamopyrgus antipodarum*)	Sediment toxicity test, 8 weeks	Embryo production (Mortality and total embryo number were also assessed)	EC_{10}: 0.19 μg/kg_{dw} (sediment) EC_{50}: 5.67 μg/kg_{dw} (sediment) LOEC: 1 μg/kg_{dw} (sediment)	(Duft et al., 2007, citing Tillman 2004)
BPA	Marine netted whelk (*Nassarius reticulatus*)	Sediment toxicity test, 4 weeks	Imposex, mortality, gland weight	NOEC: 100 μg/kg_{dw} (sediment)	(Duft et al., 2007, citing Tillman 2004)
BPA	Terrestrial isopod (*Porcellio scaber*) (Crustacea: Isopoda)	Soil toxicity test, 28 days Adult males	Reduced total ecdysteroid (20E) concentration	LOEC: 10 mg/kg_{dw} (soil)	(Lemos et al., 2009)

[a] see footnote descriptions on last page of Table.

Table 8-3. Summary of Soil and Sediment Ecotoxicology Data for High Priority TOrCs (continued).

Chemical	Test organism	Study Description	Effect	Effect Concentration	Data Source
BPA	Terrestrial isopod (*Porcellio scaber*) (Crustacea: Isopoda)	Soil toxicity test, 10 weeks Adult males	Reduced total ecdysteroid (20E) concentration	LC_{50}: 910 mg/kg_{dw} (soil)	(Lemos et al., 2009)
BPA	Terrestrial isopod (*Porcellio scaber*) (Crustacea: Isopoda)	Soil toxicity test; 16 weeks Immature, sexually undifferentiated organisms	Survival	LC_{50}: >1,000 mg/kg_{dw} (soil)	(Lemos et al., 2009)
BPA	Terrestrial isopod (*Porcellio scaber*) (Crustacea: Isopoda)	Soil toxicity test; 16 weeks Immature, sexually undifferentiated organisms	Sex ratio	LOEC: 10 mg/kg_{dw} (soil) Higher concentrations had no significant effect.	(Lemos et al., 2009)

Notes:

1 Nominal concentrations. Measured concentrations were smaller, varying from ~40-90% of nominal concentrations.

2 The test chemical is described as decabromodiphenyl either (CASRN 1163-19-5). One of the chemical names associated with this CASRN is BDE 209.

3 Nominal concentrations. Measured concentrations were generally much greater.

4 Range of toxicity values for 6 different tested topsoil types.

5 OECD Guideline for Testing of Chemicals 207, Section 2 (adopted April 4, 1984). Artificial soil deviated from the study protocol requirements in that 21% of the sand was exchanged for natural soil, promoting greater absorptive capacity. The test substance was mixed with dry soil, which is known to enable maximum absorption. Consequently, the test may have underestimated the toxicity of triclosan to worms (Samsøe-Petersen et al. 2003).

6 Wüthrich, V. Report 262956. Unpublished data. Itingen, Switzerland: RCC, Umweltchemie.

7 Vegetative vigor study: Schwab, D; L.G. Heim, Report 42620. Unpublished data. Columbia, Missouri: Analytical Bio-Chemistry Laboratories. Study conducted according to prior FDA guidelines (http://vm.cfsan.fda.gov/~dms/opa-eg11.html).

8 According to Samsøe-Petersen et al. (2003), "The result of this study is quoted by Mones as a NOEL for all parameters (shoot length, shoot and root weight) of '>424 1000 µg/kg' (presumably >424- 1000 µg/kg)" while Reiss et al. 2009 (and Reiss et al. 2001, as cited by Samsøe-Petersen et al. (2003)) state that there were no effects up to 1000 µg/kg. According to Reiss et al. (2009), effects of the vehicle blank on shoot length and shoot dry weight lend uncertainty to the results.

9 Pre-emergent study (seedling emergence and growth): J.R. Hoberg, Report 90-12-3574. Unpublished data. Wareham, Massachusetts: Springborn Laboratories. Study conducted according to prior FDA guidelines (http://vm.cfsan.fda.gov/~dms/opa-eg11.html).

10 Soil respiration was inhibited during the first 4 days of incubation but recovered after longer incubation, probably due to biodegradation of triclosan in soil.

11 Phosphatase activity was inhibited for all soils treated with triclosan (from 0.1-50 mg/kg dry soil), but declining inhibition was observed after 2 days of incubation, probably due to biodegradation of triclosan in soil.

Table 8-4. Summary of Relevant Human Toxicity Data for High Priority TOrCs.

Chemical	Toxicity Information Description	Toxicity Value	Data Source
CIP	ADI - based on sensitivity of human intestinal microflora, i.e., minimum inhibitory concentration (MIC) values for ciprofloxacin against human intestinal flora	1.6 µg/kg/day	(Schwab et al., 2005)
CIP	ADI (S-ADI derive using a safety factor)	7.1 µg/kg/day	(EPHC, 2008)
Doxycycline	ADI	3 µg/kg/day	(EPHC, 2008)
Doxycycline	ADI based on sensitivity of human intestinal microflora, i.e., based on antimicrobial sensitivity of human intestinal flora	30 µg/kg/day	(Schwab et al., 2005)
TC	ADI	30 µg/kg/day	(EPHC, 2008)
TC	ADI (group ADI for tetracycline, oxytetracycline, and chlortetracycline)	0.03 mg/kg_{bw}	(JECFA, 1998)
TC	ADI based on sensitivity of human intestinal microflora, i.e., based on antimicrobial sensitivity of human intestinal microflora	30 µg/kg/day	(Schwab et al., 2005)
TCS	ADI	75 µg/kg/day	(Snyder et al., 2008a)
TCS	Chronic population adjusted dose[1]	0.30 mg/kg/day	(U.S. EPA, 2008b; U.S. EPA, 2008c)
TCS	Threshold of toxicological concern[2]	1.5 µg/kg bw/day	(EPHC, 2008)
Cimetidine	ADI (S-ADI derive using a safety factor)	5.7 µg/kg/day	(EPHC, 2008)
Cimetidine	ADI derived from lowest therapeutic dose for reducing gastric acid secretion in adults	29 µg/kg/day	(Schwab et al., 2005)
BDE 47	RfD (critical effect: neurobehavioral effects)	0.0001 mg/kg-day	(U.S. EPA-IRIS, 2009)
BDE 99	RfD (critical effect: neurobehavioral effects)	0.0001 mg/kg-day	(U.S. EPA-IRIS, 2009)
BDE 153	RfD (critical effect: neurobehavioral effects)	0.0002 mg/kg-day	(U.S. EPA-IRIS, 2009)
BDE 209	10^{-6} cancer risk level[3]	50 µg/L	(U.S. EPA-IRIS, 2009)
BDE 209	RfD (critical effect: neurobehavioral effects)	0.007 mg/kg-day	(U.S. EPA-IRIS, 2009)
PFOA	Chronic, non-cancer RfD (critical effect: increased relative liver weight; with co-critical effects[4]) (critical effects: developmental, liver, and immune system effects)	0.000077 mg/kg/day	(Minnesota Department of Health, 2008; Minnesota Department of Health, 2009a; Minnesota Department of Health, 2009b)
PFOA	Industrial Soil Reference Value (based on RfD of 0.001 mg/kg/day)[5]	200 mg/kg	(Minnesota Department of Health, 2005)
PFOA	Residential Soil Reference Value (based on RfD of 0.001 mg/kg/day)[5]	30 mg/kg	(Minnesota Department of Health, 2005)
PFOA	Subchronic RfD (U.S. EPA-OSWER 2009) for liver effects, based on provisional short-term value (U.S. EPA 2009a,b) of 0.0004 mg/L (2009 Health Assessment)[6]	0.0002 mg/kg/day, based on a provisional short-term value of 0.0004 mg/L	(U.S. EPA-OSWER, 2009; U.S. EPA-OW, 2009)
PFOS	Chronic, non-cancer RfD (critical effects: developmental, liver, thyroid[7]) (critical effects: decreased high density lipoprotein cholesterol, decreased total triiodothyronine, increased thyroid-stimulating hormone; co-critical effects: decreased body weight and body weight gain in offspring)	0.00008 mg/kg/day	(Minnesota Department of Health, 2008; Minnesota Department of Health, 2009a; Minnesota Department of Health, 2009b)
PFOS	Industrial Soil Reference Value (based on RfD of 0.0002 mg/kg/day)[8]	40 mg/kg	(Minnesota Department of Health, 2005)
PFOS	Residential Soil Reference Value (based on RfD of 0.0002 mg/kg/day)[8]	6 mg/kg	(Minnesota Department of Health, 2005)
PFOS	Subchronic RfD (U.S. EPA-OSWER, 2009) for thyroid and plasma lipoprotein effects, based on provisional short-term value (U.S. EPA 2009a,b) of 0.0002 mg/L (2009 Health Assessment)[9]	0.00008 mg/kg/day, based on a provisional short-term value of 0.0002 mg/L	(U.S. EPA-OSWER, 2009; U.S. EPA-OW, 2009)

[a] see footnote descriptions on last page of Table.

Table 8-4. Summary of Relevant Human Toxicity Data for High Priority TOrCs (continued).

Chemical	Toxicity Information Description	Toxicity Value	Data Source
4-Cumylphenol	Threshold of toxicological concern[10]	1.5 $\mu g/kg_{bw}/day$	(EPHC, 2008)
4-tert-Octylphenol	ADI (critical effect: reproductive developmental toxicity in rats)	150 µg/kg-day	(Snyder et al., 2008a)
BPA	ADI (critical effect: developmental toxicity)	50 µg/kg-day	(Snyder et al., 2008a)
BPA	ADI[11]		Japanese government (Snyder et al., 2008a)
BPA	Oral RfD	0.05 mg/kg/day	(USGS, 2009)
BPA	Provisional tolerable daily intake (developed by the Health Canada Food Directorate)	25 µg/kg-day	(Health Canada, 2008)
BPA	RfD (critical effect: reduced mean body weight)	0.05 mg/kg/day	(U.S. EPA-IRIS, 2009)
BPA	Tolerable daily intake	0.01 mg/kg/day	(European Commission SCF, 2002)
BPA	Tolerable daily intake	16 µg/kg-day	Willhite (2007) for NSF International, as cited by (Health Canada, 2008)

Notes:

1 Calculated by dividing the 5th percentile NOEL by a safety factor of 100. This is the value used by various authorities in assessing risks associated with minor contaminants in food.

2 The highest predicted dose to which a person could be exposed over the course of a lifetime with no expected adverse health effect. For additional information, see "Dietary Risk Assessment for Triclosan, dated August 11, 2008.

3 10-6 cancer risk level (1 in one million cancer risk level, i.e., increased lifetime chance of 0.000001 in 1 (or one chance in a million) of developing cancer due to lifetime exposure to a substance.

4 Co-critical effects: increased liver weight with histopathological changes, decreased total serum cholesterol and triglycerides, developmental delays (e.g., altered body weight gain, delayed physical development, hepatocellular hypertrophy) in offspring, altered immune function.

5 Based on the Minnesota Department of Health's RfD of 0.001 mg/kg/day. Soil concentrations less than this value should provide an adequate level of protection from exposure due to direct contact with contaminated soil. This value should be considered provisional. This criterion does not address impacts to groundwater as a result of soil leaching, food chain impacts, or ecological impacts. Although carcinogenicity studies in the rat have shown that PFOA is potentially carcinogenic, the data available at the time when this criterion was proposed were not considered to be sufficient to determine the relevance to humans or for development of cancer potency values.

6 Provisional Health Advisory values are developed to provide information in response to an urgent or rapidly developing situation. They reflect reasonable, health-based hazard concentrations above which action should be taken to reduce exposure to unregulated contaminants in drinking water. They will be updated as additional information becomes available and can be evaluated.

7 Endocrine mediated effect on the thyroid.

8 Based on the Minnesota Department of Health's RfD of 0.0002 mg/kg/day. Soil concentrations less than this value should provide an adequate level of protection from exposure due to direct contact with contaminated soil. This value should be considered provisional. This criterion does not address impacts to groundwater as a result of soil leaching, food chain impacts, or ecological impacts. Although carcinogenicity studies in the rat have shown that PFOS is potentially carcinogenic, the data available at the time when this criterion was proposed were not considered to be sufficient to determine the relevance to humans or for development of cancer potency values.

9 Provisional Health Advisory values are developed to provide information in response to an urgent or rapidly developing situation. They reflect reasonable, health-based hazard concentrations above which action should be taken to reduce exposure to unregulated contaminants in drinking water. They will be updated as additional information becomes available and can be evaluated.

10 Calculated by dividing the 5th percentile NOEL by a safety factor of 100. This is the value used by various authorities in assessing risks associated with minor contaminants in food.

11 Japan convened an expert group that conducted a risk assessment for BPA. Snyder et al. (Snyder et al., 2008a) reported ADIs that corresponded to doses used in that risk assessment. This is the lowest ADI reported from that risk assessment.

Relevant soil and sediment toxicity data were found for very few of the high priority TOrCs targeted in this review. Even when relevant ecotoxicity values were identified, they were extremely limited in terms of quantity, study quality, toxicological endpoints that were investigated, and number of species and taxa that were evaluated. A significant proportion of the available studies were conducted in sediment, so the results may be less applicable to prediction of ecotoxicological effects in soil. Due to budget and time constraints, the current review of the known and potential modes of action in for these TOrCs in ecological receptors is far from comprehensive and should be expanded in future efforts. Particularly for those TOrCs (i.e., PFCs) that may be significantly leached into groundwater or surface water or that are bioavailable in soil water, in sediment pore water, or in water at the sediment-water interface, the available aquatic toxicity data based on waterborne exposures may provide information that is valuable for risk assessment of contaminants in biosolids. For some of these TOrCs, there are significant databases of aquatic toxicity information based on waterborne exposures or based on internal concentrations of contaminants. For those TOrCs for which the exposure information suggests that these may be important and relevant for risk assessment, these additional toxicity values should be gathered and reviewed.

Table 8-5 provides a summary of the general data availability with respect to the high priority TOrCs included in this study. The decision criteria used to bin the TOrC classes or subclasses is indicated, though in some cases, expert judgment was required to consider data that did not readily fit within the framework developed. Significant data gaps with respect to toxicity, particularly ecological toxicity, are evident for many of the TOrCs included in this study.

Table 8-5. Summary of Toxicity Data Availability for the High Priority TOrCs.

Chemical Class	Toxicity Data Availability	
	Human	Ecological
BFRs	Tier 0	Tier 0
PFCs and PFC Precursors	Tier 0	Tier 0
PPCPs: Antimicrobials	Tier 0	Tier 0
PPCPs: Antibiotics	Tier 2[1]	Tier 0
PPCPs: Synthetic Musks	Tier 1	Tier 0
PPCPs: Other	Tier 2[1]	Tier 0
Plasticizers	Tier 1	Tier 0
Steroidal Chemicals	Tier 2[1]	Tier 0
Surfactants	Tier 1	Tier 0

[1] The potential health effects of therapeutic use of drugs may be considered to be well-characterized due to testing required during drug development and registration. However, potential health effects of non-therapeutic exposure (e.g., longer term exposures, sub-therapeutic doses, unusual routes of exposure) and exposure of non-target populations (e.g., people for whom the use of the drugs is contraindicated) are often unknown.

Data Availability Ranking Decision Criteria ~ Human Toxicity:

Tier 0 (No Data)	Essentially no data were available of this type for this class or subclass of TOrCs, including data that could be used for modeling.
Tier 1	For the majority of TOrCs in this class or subclass, substantial animal toxicological data (both acute and chronic) are available.
Tier 2	For the majority of TOrCs in this class or subclass, substantial human toxicological data are available.
Tier 3	For the majority of TOrCs in this class or subclass, human health benchmark (HHB) values been derived. Note: In some cases, HHBs are derived simply by applying safety/uncertainty factors to the therapeutic dose. Thus, the existence of an HHB does not necessarily imply the availability of a substantial body of toxicological information useful for risk assessment of soil exposure for the general population.

Data Availability Ranking Decision Criteria ~ Ecological Toxicity:

Tier 0 (No Data)	Essentially no data were available of this type for this class or subclass of TOrCs, including data that could be used for modeling.
Tier 1	For the majority of TOrCs in this class or subclass, relevant soil or sediment ecotoxicological data representing multiple taxa are available in either sediment or soil.
Tier 2	For the majority of TOrCs in this class or subclass, relevant ecotoxicological data representing multiple taxa are available in soil systems.
Tier 3	For the majority of TOrCs in this class or subclass, ecotoxicological data are available for these TOrCs in biosolids-amended soils

CHAPTER 9.0

IMPACTS OF BIOSOLIDS-BORNE TRACE ORGANIC CHEMICALS ON SOIL MICROORGANISMS

9.1 Introduction

The presence of environmental contaminants in soils can affect soil microbial communities in measureable ways. These include alterations in community composition as measured by phylogeny, rates of growth or cellular density, and metabolic function and diversity. The microbial impacts resulting from selective pressures such as the introduction of sufficient quantities of a primary energy source (i.e., electron donors and acceptors such as sugars, fatty acids, and oxygen) are better understood than the impact of many trace organic chemicals (TOrCs). In some cases, the nature of certain classes enables some predictions of community effects. For example, pharmaceuticals and personal care products (PPCPs) that have antibiotic properties might be expected to suppress microbial growth for susceptible organisms while limiting competition for those with resistance. Data that are available with respect to these types of microbial impacts as well as knowledge gaps are addressed in this chapter.

Biosolids amendment can result in paradoxes where potentially competing processes such as nutrient and carbon addition cannot be easily contrasted with TOrC presence or toxicity. Further complications arise due to variables such as endogenous soil microbes, biosolids type, application process, climate, and soil type as well as the concentration, composition, and cumulative effects of a suite of TOrCs. Available data do not adequately address these issues and the observed effects of biosolids application such as increased microbial respiration (Banks et al., 2006), shifts in community profile (Brooks et al., 2007a; Lapen et al., 2008a), or increased nitrification and enzyme activity (Zerzghi et al., 2010) may be due to any of the above factors.

Studies evaluating the effects of six classes of TOrCs (PPCPs, brominated flame retardants [BFRs], steroidal chemicals, perfluorochemicals [PFCs], surfactants, and plasticizers) on soil microbial communities were sought and examined. Community effects were typically measured via genetic, microbiological, and biochemical assays. Phylogenetic diversity and alterations in consortia structure were measured by methods that investigate cellular composition such as fatty acid analysis (i.e., phospholipid fatty acid analysis; PLFA) as well as nucleic acid based techniques to identify primary cellular presence including 16S ribosomal ribonucleic acid (rRNA) cloning and denatured gradient gel electrophoresis (DGGE) fingerprints of 16S rRNA. Population size was characterized using molecular techniques as well as by recording bulk population density and growth. Functional pressures on substrate utilization (BIOLOG broad substrate utilization array) and transformative biochemical processes such as nitrification were also recorded for certain compounds. Quantification of these processes enables the determination of variables such as effective concentrations where a significant impact on rates is found. The impact of certain TOrCs on cellular respiration and enzyme expression and activity was also quantified.

There is no uniformity among functional and/or structural effects on soil microbes reported in the literature, which is complicated by the diversity of microbial analyses reported. This chapter focuses on synthesizing the limited data available for impacts of TOrCs on microbial composition and function to consolidate what is known and what data gaps exist regarding the potential impact of biosolids-borne TOrCs on soil microbe dynamics. Before significant experimental progress can be made in understanding the impacts of TOrCs and identifying areas of further research, methodology standardization is needed to determine appropriate baselines and metrics.

9.2 Pharmaceuticals and Personal Care Products

PPCPs are found ubiquitously in multiple environments related to the biosolids focus of this review. These include salt marshes, freshwater systems, agricultural soils, and activated sludge as well as other compartments of wastewater treatment plants (WWTPs). Due to their pervasiveness in the environment, the study of antibiotics and antimicrobials and their impact on natural soil communities and processes comprise a large portion of PPCP research. A handful of agents have been the focus of the bulk of published research, including ciprofloxacin (CIP), various tetracyclines, triclosan (TCS), and triclocarban (TCC).

Fluoroquinolone Antibiotics

Increasing aqueous concentrations of CIP (0.02 - 200 µg/mL) increased both microbial biomass and richness in salt marsh sediment microcosms, as determined by PLFA (Cordova-Kreylos and Scow, 2007). Additionally, treatment of the sediment with 0.2, 2.0, 100 and 200 µg/mL ciprofloxacin increased the proportion of sulfate-reducing bacteria in the community. At higher concentrations, bacteria possessing monoenoic fatty acids were more prevalent than those with branched fatty acids. Compared to controls, the microbial community exposed to CIP exhibited decreased ratios of the 17:0 cy/precursor stress biomarker as well as saturated and unsaturated PLFAs. The CIP-treated communities appeared less stressed than controls. The authors proposed that this increase in biomass and decrease in stress biomarker expression was due to the utilization of CIP as a carbon source. The microorganisms were derived from sediments, and it is uncertain whether analogous behavior would be seen in environments more relevant to biosolids amendment of soils.

Antimicrobial Agents

The effects of TCS and TCC on microbial communities have been studied in biofilm communities, activated sludge, WWTP influent, benthic communities and aerobic soils. In domestic sink drain biofilms, TCS-containing detergents (0.2 and 0.4% wt/v) did not affect microbial community size, but DGGE analysis identified a shift in bacterial diversity. Tetracycline- and TCS-resistant microbes were present in both the controls that had never received antibiotics or biocides and TCS-treated microcosms (McBain et al., 2003).

River biofilms were cultivated in microcosms supplemented with 10 µg/L TCS or TCC for a period of 8 weeks (Lawrence et al., 2009). Both compounds impacted the structure, composition, biomass, metabolism, and diversity of the microbial community. Compared to controls, TCC-treated communities exhibited significantly increased biofilm thickness as well as a more extensive confluent biofilm compared to TCS-treated communities. Both antimicrobials significantly reduced algal biomass, which resulted in a shift from primarily autotrophic communities to heterotrophic ones. TCC and TCS also impacted the metabolism of heterotrophic organisms. BIOLOG assays indicated that TCC suppressed select carbon substrate utilization including certain carbohydrates, carboxylic acids, and amino acids, while TCS suppressed all substrate utilization. The effects of TCC and TCS on diversity were measured via lectin-binding analyses of the biofilm's extracellular polysaccharide matrix. Both TCC- and TCS-treated communities exhibited a significant increase in *Canavalia ensiformis* binding dominance, while TCC-treated biofilms also exhibited significant increases in *Glycine max* and *Arachis hypogeae* lectin binding. The change in microbial diversity was confirmed via DGGE analysis in which banding patterns for treated communities differed from controls. In contrast to these biofilm studies, neither 1 mg/kg TCC nor TCS affected microbial activity in aerobic soil samples (Ying et al., 2007) suggesting that pronounced differences may be observed if drawing conclusions from soil-derived microbial enrichments or even biofilms rather than testing the endogenous soil consortium in its natural environment.

Regarding the effects of TCS on the activated sludge process, it was found that the extremely high doses (likely exceeding the aqueous solubility) to activated sludge-mixed liquor inhibited heterotrophic activity by 50% (Reiss et al., 2009). In this same study, TCS did not affect soil nitrification or microbial respiration at concentrations up to 2 mg/kg. In other work, 0.5 mg/L TCS on non-acclimatized biomass initially deteriorated ammonia removal and nitrification capacity (Stasinakis et al., 2007). However, after biomass acclimation, nitrification was fully recovered, and a further increase of TCS to 2 mg/L did not affect the performance of the activated sludge system. The effect of TCS on the ability of the activated sludge to remove bulk organic substrates proved minor for concentrations up to 2 mg/L, indicating that heterotrophic microorganisms are less sensitive to TCS than are nitrifiers. Additionally, increasing levels of TCS in WWTP influent had no major adverse effect on secondary and advanced wastewater treatment process such as biochemical oxygen demand (BOD) and ammonia removal (Federle et al., 2002). Batch studies revealed acclimation of microbes to TCS and potential selection or adaptation of the microbial community to biodegradation of the contaminant. The addition of 200 μg/L TCC in a continuous flow activated sludge system (retention time = 10 h) resulted in acclimation of microbes to the primary biodegradation of the chemical as evidenced by increased rates of attenuation (Gledhill, 1975).

Tetracycline Antibiotics

As a prevalent group of PPCPs in the environment, tetracyclines have been studied for their impact on freshwater sediments, agricultural soils, and WWTP processes. Though not a target of this study due to its low concentration in biosolids, several studies examining the effects of oxytetracycline (OTC) have been conducted and may provide insight into potential microbial impacts of other tetracyclines that are included in the present study (i.e., 4-epitetracycline, doxycycline, tetracycline [TC], anhydrotetracycline, minocycline). In an attempt to determine the impact of OTC on soil microbes (Kong et al., 2006), paddy soil was exposed to varying aqueous concentrations of OTC (0, 1, 5, 11, 43, 109 and 217 μM). Microbial diversity was reduced 63% at 43 μM OTC, and evenness was reduced 41% at 109 μM OTC. Diversity and evenness were based on BIOLOG substrate utilization data and calculated using the Shannon index. Microbial activity as indicated by average well color development was reduced, with a critical OTC dose found at the lowest tested concentration (1 μM) where microbial substrate utilization was reduced in excess of 50%. The combination of OTC and copper (20 μM Cu + 5 μM OTC, 20 μM Cu + 11 μM OTC) had an additive effect that significantly decreased microbial diversity and utilization of carboxylic acids and carbohydrates when compared to OTC or copper alone. However, a study of agricultural soils found that OTC did not affect microbial activity, as determined by basal respiration and dehydrogenase activity, at concentrations up to 1000 µg/g (Thiele-Bruhn and Beck, 2005) While incubation periods of 4 hours did not affect microbial growth, a longer incubation period in the presence of OTC (24 h) did inhibit growth. During these longer duration experiments, OTC significantly reduced bacterial numbers, with effects persisting for several weeks after the antibiotic was no longer detectable in the soil (Thiele-Bruhn and Beck, 2005).

In activated sludge, EC_{50} (50% effective concentration) values of OTC, TC, and chlortetracycline (CTC; a related tetracycline-derived antibiotic) were determined (Loke et al., 2002). The EC_{50} values of TC and OTC were both 0.08 mg/L, and the EC_{50} of CTC was 0.03 mg/L. The minimum concentration required to inhibit 50% of growth were also determined for the above tetracyclines tested against tetracycline-sensitive Pseudomonads, an important class of soil microbes believed to be involved in the biotransformation of a wide variety of xenobiotic compounds. These values were found to be 2.0 mg/L for TC, 0.5 mg/L for CTC, and 1.0 mg/L for OTC. Degradation products derived from parent compounds were also tested in activated sludge and soil bacteria to determine their comparative toxicity. Several of these degradation products, including 4-epi-anhydro-tetracycline hydrochloride, 5a,6-anhydrochlorotetracycline hydrochloride, and 4-epi-oxytetracycline exhibited similar effects to the parent compounds with slightly lower toxicity. In another study, TC and other broad spectrum antibiotics exhibited increased toxicity (lower EC_{50}) to activated sludge aerobes compared to narrow spectrum antibiotics (Halling-Sorensen, 2001). Additionally, microbes were isolated from the sludge that were resistant to the highest level of TC tested (8 mg/L).

Understanding the distribution of antibiotic resistance genes (ARGs) also provides useful information regarding the structure and composition of soil microbial communities as well as the propagation of this resistance. The presence of tetracycline-resistance and other ARGs have been studied in systems receiving both animal wastes and municipal biosolids. Such studies are more generally related to horizontal gene transfer and microbial selection in impacted systems. The identified studies focus primarily on aquatic systems indicating a need for research pertaining to ARGs in soil environments.

In one comparison study, the presence / absence of five tetracycline and four sulfonamide ARGs were determined in river sediment, dairy lagoon water, irrigation ditch water, a wastewater recycling plant, and two drinking water treatment plants (Pruden et al., 2006). River sediment samples collected at multiple time points were pooled and ARG levels normalized to total 16S rRNA gene levels. Except for the resistance gene sul(II), there were no significant differences in ARG concentrations among the collection sites. Non-normalized ARG data did show a significant different between several river collection sites. The overall trend in ARG concentration correlated with anthropogenic inputs as: river sediment < ditch water < lagoon water. Tet(W) and tet(O) were also detected in various samples collected from drinking water treatment plants and a wastewater recycling plant. In another study, the ARGs tet(O), tet(Q), tet(W), tet(M), tetB(P), tet(S), tet(T), and otrA were identified in total deoxyribonulcelic acid (DNA) extracted from water of two waste lagoons near swine production facilities, and tet(M) was found in groundwater near the facilities (Chee-Sanford et al., 2001; Chee-Sanford et al., 2009). Admittedly, un-digested manure and biosolids differ greatly in their microbial phylogeny and their concentrations of TOrCs such as antibiotics; however, the leaching of ARGs into the environment from manure suggests that leaching may also occur in biosolids-amended soils.

The presence or absence of ten tetracycline ARGs (tet(A)-(E), tet(G), tet(M), tet(O), tet(Q), tet(S)) was determined for two WWTPs and two freshwater lakes (Auerbach et al., 2007). Only tet(A) was identified in the two lakes, while multiple resistance genes were present in all WWTP samples indicating selection for resistance in these systems. Notably, biosolids derived from the two WWTPs contained varying levels of ARGs.

In a sandy loam injected with liquid biosolids, both the soil and the biosolids were analyzed for levels of antibiotic-resistant bacteria over a period of 15 months (Brooks et al., 2007b). Soil and biosolids-borne organisms were assayed against four antibiotics: ampicillin (32 μg/mL), cephalothin (32 μg/mL), ciprofloxacin (4 μg/mL), and tetracycline (16 μg/mL). In amended soils, the number of antibiotic-resistant bacteria did not significantly change over the 15 months and were not significantly different from un-amended control sites. Unfortunately, this study did not quantify the presence of antibiotics in the biosolids or the soil, which would have been beneficial to interpreting their results.

9.3 Brominated Flame Retardants and Steroidal Chemicals

Both BFRs and steroidal chemicals are of particular concern due to their widespread presence in the environment. Limited data exist, however, to describe the impact of these chemicals on soil microbial growth and diversity.

In a toxicity study of several BFRs (Sverdrup et al., 2006), short-chain chloroparaffins (i.e., polychlorinated *n*-alkanes) repressed soil nitrification with an estimated EC_{10} value (concentration where 10% repression effect is observed) of 570 mg/kg, the highest concentration tested in this study. Tetrabromobisphenol A (TBBPA) had a more pronounced effect on nitrate production with a calculated EC_{10} of 295 (range 210-390) mg/kg or about half that recorded for chloroparaffins. However, considering the concentrations of chloroparaffins and TBBPA found in municipal biosolids (1.8-93.1 and 0.002-0.6 mg/kg respectively; Table 4-1) it is questionable whether toxicity is likely to be observed in biosolids-amended soils. In contrast, no observable effect on nitrification rates was reported in the presence of decabromodiphenyl ether (BDE 209) at levels up to 1000 mg/kg soil dry weight.

Steroidal chemicals as environmental contaminants may affect microbial metabolic regulation in a variety of ways. No data were available for the compounds included in this study. However, trenbolone, a hormone used in cattle production, had little impact on the total gene pool and phylogeny of soil extractable bacterial community structure as determined with 16S rRNA gene analysis (Radl et al., 2005). Trenbolone did significantly affect community function. In soil/water microcosms containing 15 μg/L trenbolone (1.5 g water / 3.2 g sediments), a 50% reduction of *N*-acetyl-glucosaminidase activity was observed. Decreases in leucine-aminopeptidase, and b-glucosidase activity, and an overall decrease in substrate utilization potential as determined by the BIOLOG broad substrate assay were also observed.

9.4 Perfluorochemicals and Surfactants

In un-amended soil microcosms, 90 mg/kg 4-nonylphenol (4-NP) did not affect the population distribution of a community of soil invertebrates (Domene et al., In press). A 4-NP concentration of 270 mg/kg significantly affected species abundance within the community, but the effect was temporary and disappeared within 112 days suggesting biodegradation of 4-NP.

Though significant data exists regarding PFCs and surfactants as environmental contaminants, effects on soil microbes remain largely unknown. The absence of available data indicates a need for study in this area.

9.5 Plasticizers

Plasticizers are used in a wide variety of industries and are yet another class of TOrCs whose impact on microbial ecology is under study. Data relating to bisphenol A (BPA) impacts in soils was not found; however it is possible that analogies can be drawn from other analogous compounds of concern. The phthalate-degrading bacteria *Comamonas acidovorans* was isolated from soil treated with 0.1 to 100 mg/g diethyl phthalate (DEP) and di (2-ethyl hexyl) phthalate (DEHP) (Cartwright et al., 2000). DEP (0.1 mg/g) did not affect structural or functional diversity of the microbial community as determined via total number of bacteria, fatty acid analysis, and BIOLOG substrate utilization array. However, the number of Pseudomonads in DEP-treated soil was significantly higher than for controls at this concentration. DEP concentrations in excess of 1 mg/g resulted in a 47% reduction in total culturable bacteria and a 62% reduction in Pseudomonads within 1 day of exposure. In contrast, DEHP had no effect on diversity or population size at similar concentrations.

Metabolites of plasticizers can also be environmental contaminants. Pure culture analyses (Nalli et al., 2006) have shown that concentrations of 1, 1.5, and 2 mmol/L of 2-ethylhexanol – a metabolite of DEHP – inhibited degradation of adipic acid by *Rhodococcus rhodochrous*. Itself a metabolite, adipic acid was capable of supporting growth for *R. rhodochrous* at concentrations less that 1 g/L; however, growth and metabolic activity were reduced when compared to cells grown on hexadecane. To determine the effect of 2-ethylhexanol on metabolism, alcohol dehydrogenase activity was measured via conversion of 2-ethylhexanol to 2-ethylhexanoic acid. Rates of biotransformation decreased with increasing concentrations of adipic acid. Similar decreases in enzyme activity were found when the presence of 2-ethylhexanol was equal to or greater than 1.5 mmol/L.

9.6 Conclusions

While published data relating to the microbial impacts of PFCs and surfactants are not currently available, studies relating to the remaining TOrC classes demonstrate a variety of possible effects on microbial structure and function. Data pertaining to the microbial impacts of most TOrCs included in this study were not available, and in some cases (plasticizers and steroidal chemicals), the only available data within the selected class were for TOrCs that have been excluded from this study for various reasons. Moreover, few studies have actually examined microbial impacts of TOrCs in soils, much less biosolids-amended soils. Due to the diversity of TOrCs and uncertainty associated with extrapolating effects from liquid to a complex biosolids matrix, it is difficult to make overarching conclusions about the microbial impacts of TOrCs in biosolids-amended soils. Effects such as suppression of microbial diversity and nitrifying activity would likely have an adverse effect on both TOrC attenuation and crucial microbially-driven nutrient cycling processes.

Though no microbial impact data were available for BPA, more well-studied plasticizers generally did not affect the functional diversity of soil communities but may select for increased abundance of certain microbial populations. Notably, certain plasticizer metabolites reduce the activity of alcohol dehydrogenase, an enzyme involved in the degradation of alcohol moieties (i.e., conversion of 2-ethylhexanol to 2-ethylhexanoic acid) (Nalli et al., 2006). Neither hormones nor BFRs appeared to affect microbial structure as determined by total gene pool, phylogenetic changes in composition, or community structure and distribution. However, certain BFRs impact functional activity as determined by soil nitrification rates (Sverdrup et al., 2006). Similarly, the hormone trenbolone reduced the ability of microbes to utilize certain substrates when compared to controls as quantified by a broad substrate assay (BIOLOG) as well as reduced activity for target enzymes of interest (Radl et al., 2005). These data suggest that endocrine-disrupting compounds may play a more prominent role in interfering with microbial function and metabolism than in altering overall community structure.

PPCPs have been the focus of much of the microbial impacts research to date. Effects vary with each compound, and sometimes between different studies that utilize the same compound. The presence of the antibiotic CIP in salt marsh sediments increased microbial biomass, richness, and the sizes of certain bacterial populations such as sulfate reducers and those high in certain fatty acids rather than apply direct selective pressure that would decrease diversity and suppress populations as might be expected for biocides (Cordova-Kreylos and Scow, 2007). One study investigating the impact of tetracyclines reported reduced microbial diversity, evenness or population distribution, enzyme activity, and substrate utilization (Kong et al., 2006), while another indicated no functional effects as quantified by metabolism and enzyme activity but did demonstrate a reduction in total bacterial biomass (Thiele-Bruhn and Beck, 2005). The antimicrobials TCS and TCC exhibited a wide variety of effects on river biofilms, including changes to microbial communities structure, metabolism, and diversity (Lawrence et al., 2009). Conversely, these same compounds in aerobic soil studies exhibited no effect on soil microbial communities (Ying et al., 2007).

The identified studies represent the majority of published work into how TOrC presence and concentration might affect microbial structure and function. Since anticipated attenuation phenomena for these TOrCs are likely a result of microbiological processes, a better understanding of potential selective pressures would better address risks and loads associated with the land application of biosolids. It is also possible that correlations could be made between community composition or enzyme presence and biodegradation potential. Little information was available for many of the TOrCs, particularly PFCs, surfactants, steroidal chemicals, and plasticizers, indicating a need for research into these classes of TOrCs and their potential effects on soil microbes. There is an obvious dependence upon both compound-specific attributes and the surrounding environmental system suggesting the need for independent study for many of the biosolids-borne TOrCs included in this study.

The contents of this chapter demonstrate a clear need for additional research pertaining to the biosolids-mediated mobilization and/or transmission of ARGs in the environment. Data obtained from aquatic systems and systems using un-digested animal manure indicate that the spread of ARGs is possible. However, whether the development of resistance in environmental microbes is due to selective pressures caused by low concentrations of antibiotics in biosolids or to horizontal gene transfer among microbes remains unclear. Better understanding the mechanisms of ARG mobilization and antibiotic resistance and survival in impacted soils will enable the development of preventative measures to curb the spread of antibiotic resistance in the environment which is a topic of immense societal concern.

However, before research commences to fill the identified TOrC data gaps, several issues must first be addressed. Clearly, there exist multiple evaluative tools that are used for microbial impact analysis. One important issue that must be addressed pertains to whether parameters such as microbial community composition, diversity, species density, total biomass, metabolic activity, or antibiotic resistance are equally important or equally valid indicators of ecological impacts relating to biosolids amendment. Should these screens fill the specified TOrC data gaps and identify potentially deleterious effects instrumental for performing risk analysis? Without a standardization and correlation to evaluative tools, the extrapolation of these effects towards applicable risk assessment dealing with biosolids-amended soils is challenging. Another issue when evaluating the effects of biosolids-borne TOrCs on soil microbial communities is the effect that biosolids alone have on soil microbes. Without this baseline, it is difficult to discern the true effects of TOrCs in these biosolids-amended soils. A possible complication is that biosolids are a byproduct of a local environment that must consider both input and wastewater treatment technologies. One solution would be the development of a well-characterized reference biosolid, which can be used to elucidate impacts on soil communities in the absence of TOrCs.

Similarly, the impact of the presence of one TOrC on the biodegradation of another TOrC in biosolids-amended soils is not well understood. Such impacts could be influenced by shifts in community and enzyme composition or substrate competition. Yet another issue relevant to the interpretation of microbial impact data is the question of key parameters or characteristics of a "healthy" soil. Without such a definition, microbial data such as shifts in population size or composition, diversity, and/or functionality are insufficient for determining the true impact of a TOrC on soil health. The standardization of both evaluative tools and the impacts of reference biosolids on soil communities would benefit efforts to characterize a "healthy" soil for the purpose of then identifying adverse impacts associated with biosolids deployment.

Table 9-1 provides a summary of the microbial impact data availability with respect to the high priority TOrCs included in this study. The decision criteria used to bin the TOrC classes or subclasses are provided, though in some cases, expert judgment was required to consider data that did not readily fit within the framework developed. As is clear from the table, significant data gaps with respect to microbial impacts are evident for many of the TOrCs included in this study.

Table 9-1. Summary of Microbial Impacts Data Availability for the High Priority TOrCs.

Chemical Class	Data Availability
BFRs	Tier 0
PFCs and PFC Precursors	Tier 0
PPCPs: Antimicrobials	Tier 1
PPCPs: Antibiotics	Tier 1
PPCPs: Synthetic Musks	Tier 0
PPCPs: Other	Tier 0
Plasticizers	Tier 0
Steroidal Chemicals	Tier 0
Surfactants	Tier 0

Data Availability Ranking Decision Criteria:

Tier 0 (No Data)	Essentially no data were available of this type for this class or subclass of TOrCs, including data that could be used for modeling.
Tier 1	For the majority of TOrCs in this class or subclass, ecotoxicological data were available for microorganisms.
Tier 2	For the majority of TOrCs in this class or subclass, ecotoxicological data were available for microorganisms in soil systems.
Tier 3	For the majority of TOrCs in this class or subclass, ecotoxicological data were available for microorganisms in biosolids-amended soils.

CHAPTER 10.0

SUMMARY AND CONCLUSIONS

10.1 Overview

The land application of biosolids has a long history of demonstrated beneficial use and minimal environmental and human health risks when conducted in accordance with existing regulations. However, the presence of trace organic chemicals (TOrCs) in municipal biosolids destined for land application has received increasing attention by the public and regulatory community in recent years. While concerns about so-called "emerging" organic contaminants in biosolids, such as pharmaceutical and personal care products (PPCPs) or perfluorochemicals (PFCs), have become more pressing, many of these TOrCs have likely been present in biosolids for decades. The main objective of this study was to review existing literature and identify data gaps that limit our ability to assess potential adverse environmental and human health impacts of biosolids-borne TOrCs in soils.

To help address the question of whether the presence of TOrCs in biosolids pose a significant risk to ecological and human health following land application, the overall goals of this study were to: 1) identify TOrCs of greatest concern and 2) determine data gaps for conducting ecological and human health risk assessments that could ultimately be used to support risk management decisions. An intended outcome of this study was to also help lay the groundwork for developing future research priorities. To address these goals, the completion of three primary objectives was sought. The first objective was to determine the TOrCs of greatest concern with respect to the land application of biosolids. A second objective was to evaluate the various quantitative risk assessment approaches applicable to biosolids-borne TOrCs and identify the most important parameters for conducting ecological risk assessments and the techniques currently available for obtaining the parameter values. The third objective was to conduct a comprehensive literature review of the identified TOrCs of greatest concern and identify relevant data on fate, transport, biotransfer from soil to plants and animals, and toxicity in the terrestrial environment with the end goal of identifying the scientific data gaps for the parameters most important for conducting terrestrial risk assessments.

10.2 TOrC Selection and Prioritization

Occurrence data from two national surveys of sewage sludge were used conducted to develop a list of TOrCs whose presence in biosolids may result in unacceptable risks to human health and the environment (Table 10-1). High priority status was assigned to TOrCs present at relatively high concentrations in biosolids or TOrCs known to be toxic or bioaccumulative in aquatic systems. The list may change as new data become available. Thus, the list of high priority TOrCs should be thought of as an evolving list, and chemicals can be added or deleted as new knowledge is gained.

Table 10-1. Trace Organic Chemicals Included in this Study.

Chemical(s)	CASRN	Chemical Class (Subclass)[a]	Use
		High Priority	
BDE 28	41318-75-6	BFRs	Fire Retardant
BDE 47	5436-43-1	BFRs	Fire Retardant
BDE 85	182346-21-0	BFRs	Fire Retardant
BDE 99	60348-60-9	BFRs	Fire Retardant
BDE 100	189084-64-8	BFRs	Fire Retardant
BDE 138	182677-30-1	BFRs	Fire Retardant
BDE 153	68631-49-2	BFRs	Fire Retardant
BDE 154	207122-15-4	BFRs	Fire Retardant
BDE 183	207122-16-5	BFRs	Fire Retardant
BDE 209	1163-19-5	BFRs	Fire Retardant
Dimethyl TBBPA	37853-61-5	BFRs	Fire Retardant Metabolite
HBCD isomers	25637-99-4	BFRs	Fire Retardant
TBBPA	79-94-7	BFRs	Fire Retardant
10:2/12:2diPAPs	NA	PFCs and Precursors	Surface Coatings
10:2diPAPs	NA	PFCs and Precursors	Surface Coatings
6:2/8:2diPAPs	NA	PFCs and Precursors	Surface Coatings
6:2diPAPs	NA	PFCs and Precursors	Surface Coatings
8:2/10:2diPAPs	NA	PFCs and Precursors	Surface Coatings
8:2diPAPs	NA	PFCs and Precursors	Surface Coatings
FOSA	754-91-6	PFCs and Precursors	Surface Coatings
FOSAA	NA	PFCs and Precursors	Surface Coatings
N-EtFOSAA	NA	PFCs and Precursors	Surface Coatings
N-MeFOSAA	NA	PFCs and Precursors	Surface Coatings
PFDA	335-76-2	PFCs and Precursors	Surface Coatings
PFDoDA	307-55-1	PFCs and Precursors	Surface Coatings
PFDS	335-77-3	PFCs and Precursors	Surface Coatings
PFHpA	375-85-9	PFCs and Precursors	Surface Coatings
PFHxA	307-24-4	PFCs and Precursors	Surface Coatings
PFHxS	355-46-4	PFCs and Precursors	Surface Coatings
PFNA	375-95-1	PFCs and Precursors	Surface Coatings
PFOA	335-67-1	PFCs and Precursors	Surface Coatings
PFOS	1763-23-1	PFCs and Precursors	Surface Coatings
PFTeDA	376-06-7	PFCs and Precursors	Surface Coatings
PFTriDA	72629-94-8	PFCs and Precursors	Surface Coatings
PFUnDA	2058-94-8	PFCs and Precursors	Surface Coatings
Bisphenol A (BPA)	80-05-7	Plasticizers	Plasticizer
4-Epitetracycline	23313-80-6	PPCPs (Antibiotics)	Antibiotic
Ciprofloxacin (CIP)	85721-33-1	PPCPs (Antibiotics)	Antibiotic
Doxycycline	564-25-0	PPCPs (Antibiotics)	Antibiotic
Miconazole	22916-47-8	PPCPs (Antibiotics)	Antifungal
Ofloxacin	82419-36-1	PPCPs (Antibiotics)	Antibiotic
Tetracycline	60-54-8	PPCPs (Antibiotics)	Antibiotic
Triclocarban (TCC)	101-20-2	PPCPs (Antimicrobials)	Antimicrobial
Triclosan (TCS)	3380-34-5	PPCPs (Antimicrobials)	Antimicrobial
Galaxolide (HHCB)	80450-66-4	PPCPs (Musks)	Fragrance material
Tonalide (AHTN)	21145-77-7	PPCPs (Musks)	Fragrance material
Cimetidine	51481-61-9	PPCPs (Other)	Antacid
17α-Ethinyl estradiol (EE2)	57-63-6	Steroidal Chemicals	Synthetic hormone
Mestranol (MeEE2)	72-33-3	Steroidal Chemicals	Synthetic hormone
4-Cumylphenol	599-64-4	Surfactants	Detergent Metabolite
4-tert-octyl phenol	140-66-9	Surfactants	Detergent Metabolite

[a] For the purposes of data gap analysis, PPCPs considered high priority were further subclassified depending on their uses.

Table 10-1. Trace Organic Chemicals Included in this Study (continued).

Chemical(s)	CASRN	Chemical Class (Subclass)[a]	Use
Low Priority			
N-alkanes (polychlorinated)	NA	Aliphatics	Flame retardant
Polydimethylsiloxane (PDMS)	9016-00-6	Aliphatics	Organosilicone polymer
Polyorganosiloxanes	NA	Aliphatics	Organosilicone polymer
Propene (trichloro)	96-19-5	Aliphatics	Herbicide intermediate
Dibutyltin	1002-53-5	Organotins	Anti-fouling agent
Monobutyltin	2406-65-7	Organotins	Heat stabilizer/ anti-fouling agent
Tributyltin	688-73-3	Organotins	Anti-fouling Agent
Hexachlorophene (HCP)	70-30-4	Phenols	Disinfectant
Hydroquinone	123-31-9	Phenols	Photographic developing
Cresyldiphenyl phosphate	26444-49-5	Phosphate Esters	Plasticizer/flame retardant
Acetyl Cedrene	125783-65-5	PPCPs	Fragrance Material
Azithromycin	83905-01-5	PPCPs	Antibiotic
BLS	NA	PPCPs	Fluorescent whitening agent
DAS 1	16090-02-1	PPCPs	Fluorescent whitening agent
Diphenhydramine	58-73-1	PPCPs	Antihistamine
Diphenyl Ether	101-84-8	PPCPs	Fragrance material
DSBP	38775-22-3	PPCPs	Fluorescent whitening agent
Galaxolide lactone (HHCB-lactone)	NA	PPCPs	Fragrance material metabolite
Hexyl salicylate	6259-76-3	PPCPs	Fragrance material
Hexylcinnamic aldehyde (α)	101-86-0	PPCPs	Fragrance material
Ibuprofen	15687-27-1	PPCPs	Analgesic
Iso-E-Super (OTNE)	54464-57-2	PPCPs	Fragrance material
Methyl ionone (gamma)	127-51-5	PPCPs	Fragrance material
Minocycline	10118-90-8	PPCPs	Antibiotic
Musk Ketone (MK)	81-14-1	PPCPs	Fragrance material
Phantolide (AHMI)	15323-35-0	PPCPs	Fragrance material
Sulfanilamide	63-74-1	PPCPs	Antibiotic
Thiabendazole	148-79-8	PPCPs	Anthelminitic
Traseolide (ATII)	68857-95-4	PPCPs	Fragrance material
17α-Dihydroequilin	651-55-8	Steroidal Chemicals	Steroid hormone
17α-Estradiol	57-91-0	Steroidal Chemicals	Steroid hormone
17β-Estradiol (E2)	50-28-2	Steroidal Chemicals	Steroid hormone
Androstenedione	63-05-8	Steroidal Chemicals	Steroid hormone
Androsterone	53-41-8	Steroidal Chemicals	Steroid hormone
Equilenin	517-09-9	Steroidal Chemicals	Steroid hormone
Equilin	474-86-2	Steroidal Chemicals	Steroid hormone
Estriol (E3)	50-27-1	Steroidal Chemicals	Steroid hormone
Estrone (E1)	53-16-7	Steroidal Chemicals	Steroid hormone
Etiocholanolone	53-42-9	Steroidal Chemicals	Androgen metabolite
Norethindrone	68-22-4	Steroidal Chemicals	Synthetic hormone
Norgestrel	6533-00-2	Steroidal Chemicals	Synthetic hormone
Progesterone	57-83-0	Steroidal Chemicals	Steroid hormone
Testosterone	58-22-0	Steroidal Chemicals	Steroid hormone
β-Estradiol-3-benzoate	50-50-0	Steroidal Chemicals	Synthetic hormone
$C_{10}EO_x$ (Alcohol Ethoxylates)	74432-13-6 (AEOs)	Surfactants	Surfactant
$C_{11}DEA$ (Coconut Diethanol Amide)	68603-42-9	Surfactants	Surfactant
$C_{12}EO_x$ (Alcohol Ethoxylates)	NA	Surfactants	Surfactant
$C_{13}DEA$ (Coconut Diethanol Amide)	68603-42-9	Surfactants	Surfactant
$C_{14}EO_x$ (Alcohol Ethoxylates)	68154-96-1 ($C_{14-18}EO_4$)	Surfactants	Surfactant
$C_{15}DEA$ (Coconut Diethanol Amide)	68603-42-9	Surfactants	Surfactant
$C_{16}EO_x$ (Alcohol Ethoxylates)	68154-96-1	Surfactants	Surfactant
$C_{17}DEA$ (Coconut Diethanol Amide)	68603-42-9	Surfactants	Surfactant
$C_{18}EO_x$ (Alcohol Ethoxylates)	68154-96-1	Surfactants	Surfactant
Poly(ethylene glycol)s	25322-68-3	Surfactants	Polymer

[a] For the purposes of data gap analysis, PPCPs considered high priority were further subclassified depending on their uses.

10.3 Risk Assessment Modeling

An evaluation of risk assessment models was conducted to identify: 1) parameters of most importance for conducting ecological risk assessments, 2) available methods for filling the data gaps, and 3) future needs for model improvements. An intent of this effort was to help guide the data gap analysis.

Risk assessment models are used to estimate contaminant exposure and inputs to the food chain transfer models. The transfer models typically include the uptake of TOrCs by plants grown in amended fields, accumulation by fruits and vegetables, and uptake by beef and dairy cattle that consume forage and silage grown on the biosolids-amended fields. Exposure estimates are also compared to the following ecological endpoints: 1) fish, aquatic invertebrates, aquatic plants, amphibians, aquatic community and sediment biota in the farm pond; 2) soil invertebrates and plants in the agricultural field; and 3) mammals and birds in contact with the agricultural field and farm pond. Due to the large number of potential receptors, an ecological effects assessment typically focuses on a small number of indicator organisms representative of the most exposed or the most sensitive species. Other data needed for risk assessment modeling include: chemical properties such as water solubility, vapor pressure, dissociation constants (pK_a), and octanol-water partitioning coefficients (K_{ow}; where appropriate); volatilization and degradation rates; organic carbon normalized solid-water partition coefficients ($K_{oc;}$ where appropriate) and soil-water partition coefficients (K_d); bioconcentration and bioaccumulation factors (BAFs) for ecological assessments; and biotransfer factors for human health assessments.

At present, all of the methods for predicting biouptake require a K_{ow} value. Most relationships between K_{ow} and biouptake were developed for hydrophobic organic chemicals, but many of the TOrCs included in the present study are not strongly hydrophobic. Various mathematical relationships have also been developed for predicting K_{oc} and K_d from a K_{ow} value, but these relationships are highly dependent on the chemical class (structural similarities). Terrestrial prey BAFs are generally not available and suitable relationships have not been established. Thus, a BAF of 1 is generally assumed for terrestrial prey, though higher BAFs may be possible if significant biomagnification occurs. In the absence of BAFs, small mammal BAFs are used for all terrestrial vertebrate prey and earthworm BAFs are used for all terrestrial invertebrate prey. Volatilization can be estimated from chemical properties using the U.S. EPA EPIWIN computer program. Water solubility and vapor pressure values can also be used to predict the Henry's Law constant. Degradation rates (biodegradation, hydrolysis, and photolysis) are difficult to predict and thus are best measured. If no empirical values are available, a default value of zero (no degradation) is assumed.

The minimum data set required to run U.S. EPA's risk model is presented in Table 10-2. A rigorous sensitivity analysis is needed to quantify the effect of changes in model parameters on the model outcome. Minimum data set parameters should: 1) be produced using accepted and appropriate analytical techniques, published in peer reviewed studies, or reports; or 2) be appropriately estimated using U.S. EPA-approved or other peer reviewed methods. The minimum data set could be refined by establishing a screening model methodology focusing on a specific subset of exposure pathways considered relevant to the type of chemical being evaluated (e.g., bioaccumulative chemicals). However, such a methodology does not currently exist. Thus, the minimum data set focuses on parameters currently required to run the risk assessment model. The table also includes references to *in silico* models (e.g., EPI Suite, SPARC) that could be used to estimate model parameters. Further investigation is needed, though, to determine whether these estimation techniques could be used for the TOrCs identified in this review. Direct measurement of these parameters is preferred, so acceptable test methods are also included in the table.

Table 10-2. Minimum Data Set Required for the U.S. EPA Risk Assessment.

Parameter	Module(s)	Test Methods	Estimation Techniques
Health benchmark	Human risk	Cancer potency factors, ingestion reference doses	Surrogate chemical or most toxic chemical in class
Ecological benchmark	Ecological risk	Water quality criteria, soil quality criteria, lowest affect dose for population endpoint	Surrogate chemical or estimation programs like ECOSAR for aquatic life
Molecular weight (and chemical structure)	Source, Surface water	-	None
Partition coefficients	Multiple	OPPTS 835.1220 (OECD 106) OPPTS 830.7550 (OECD 107)	EPI Suite, SPARC, and established estimation equations
Water Solubility	Source, Water modules	OPPTS 830.7840 (OECD 105)	EPI Suite, SPARC
Critical pressure	Source	-	EPI Suite, SPARC
Critical temperature	Source	-	EPI Suite
Boiling point	Source	OPPTS 830.7220 (OECD 103)	EPI Suite, SPARC
Vapor pressure coefficients	Source	OPPTS 830.7950 (OECD 104)	EPI Suite, SPARC
Henry's Law Constant	Multiple	-	EPI Suite and established estimation equations
Diffusivity in air	Source	-	SPARC
Diffusion coefficient in water	Source, Groundwater	-	SPARC
Ionization equilibrium constant (requires acid base designation)	Multiple	OPPTS 830.7370 (OECD 112)	SPARC
Soil degradation rate*	Watershed, Source	OPPTS 835.3110 (OECD 301)	EPI Suite
Surface water degradation rate*	Surface water	OPPTS 835.4100 (OECD 307)	EPI Suite
Groundwater degradation rate*	Groundwater	OPPTS 835.6100	EPI Suite
Bioconcentration factors	Aquatic food web	OPPTS 850.1850 OPPTS 850.1730 (OECD 305)	EPI Suite
Bioaccumulation factors	Terrestrial food web	OPPTS 850.4800 OPPTS 850.6200 (OECD 207)	Methods available for plants (BTFs) and worms, but not well developed for other prey
Biotransfer factors	Farm food chain	OPPTS 870.8320 OPPTS 870.8340	Available methods for plant uptake, beef, and dairy

* An overall media-specific degradation rate is typically a function of degradation rate associated with specific biotic and abiotic processes such as biodegradation (aerobic and anaerobic), hydrolysis, photolysis, etc. Note that, for screening purposes, degradation rates are sometimes assumed to be zero or very low (using a highly persistent organic chemical as a surrogate) to support a conservative model simulation.

Risk assessment modeling is an iterative process. As new knowledge is obtained, the risk assessment assumptions and model formulations need to be reevaluated. Based on our current knowledge of the fate and transport of TOrCs in biosolids-amended soil, several new model formulations are proposed to better describe these processes. The new model formulations include better descriptions of: 1) the sorption of ionogenic TOrCs in soil; 2) kinetically controlled sorption of TOrCs in soil; 3) the sorption of TOrCs to colloidal material involved in facilitated transport; 4) kinetic degradation of TOrCs beyond a simple first-order loss in soil; and 5) the incorporation of biotransformation of TOrCs in plants. Perhaps most importantly, the current risk models need to be verified with field validation studies. A verification exercise should include models for predicting biosolids concentrations as well as exposure concentrations in the terrestrial environment.

10.4 Occurrence of Biosolids-Borne Trace Organic Chemicals

The process used to identify and prioritize TOrCs for consideration relied heavily on detection and quantitation in municipal sewage sludge or biosolids, and substantial occurrence data were available for nearly all of the TOrCs targeted in this study. However, for some TOrCs, notably the PFCs and PFC precursors, a substantial occurrence data base is lacking. The two national surveys employed as primary sources of occurrence data (the U.S. Geological Survery (USGS) survey and the U.S. EPA's Targeted National Sewage Sludge Survey (TNSSS)) were fairly comprehensive. The TNSSS had a substantial sample size, whereas the USGS survey was more limited in scope. Broad surveys, such as the TNSSS, are needed to ensure that representative concentrations are used in risk assessments.

During the initial data collection effort, it became apparent that this study would be limited by what has already been detected in biosolids. While this was a necessary limitation of the scope of the study, it suggests that an effort to identify TOrCs that *might potentially occur* in biosolids (and therefore *might potentially pose a risk* to humans and the environment) is needed. Such a WERF-sponsored effort was recently conducted of household chemicals in wastewater, and a similar effort is currently underway with respect to assessing impacts of wastewater treatment plant (WWTP) effluents on receiving water bodies. An effort specific to biosolids may identify TOrCs for which analytical methods should be developed. The current study is limited by what has already been measured, rather than considering what should be measured.

Data on the occurrence of biosolids-borne TOrCs in biosolids-amended soils were also sought. Not surprisingly, several of the targeted TOrCs were identified and quantified in soils amended with biosolids (either as part of an experimental plot or through normal agricultural practices). However, many of the identified studies lacked clear data as to when biosolids were last applied and/or the concentrations of the TOrCs present in the applied biosolids. Such data are *crucial* for interpretation of the levels observed, and thus there is a clear need for additional controlled field studies. Studies of this type would address the persistence and mobility of biosolids-borne TOrCs, rather than simply occurrence. These studies should employ sufficient replicates and sampling frequencies to enable meaningful interpretation of trends observed over time.

Table 10-3 provides a summary of the general occurrence data availability for the high priority TOrCs included in this study. This evaluation was limited to the availability of data for TOrCs in sewage sludge and biosolids, and thus does not include an evaluation of the data availability for TOrC occurrence in biosolids-amended soils. The decision criteria used to bin the TOrC classes or subclasses are provided, though in some cases, expert judgment was required to consider data that did not readily fit within the framework developed. Given the selection and prioritization process for inclusion of TOrCs in this study, the general availability of occurrence data for the high priority TOrCs is judged to be quite high.

Table 10-3. Summary of Occurrence Data Availability for the High Priority TOrCs.

Chemical Class	Data Availability
BFRs	Tier 3
PFCs and PFC Precursors	Tier 1
PPCPs: Antimicrobials	Tier 3
PPCPs: Antibiotics	Tier 3
PPCPs: Synthetic Musks	Tier 3
PPCPs: Other	Tier 3
Plasticizers	Tier 3
Steroidal Chemicals	Tier 3
Surfactants	Tier 3

Data Availability Ranking Decision Criteria ~ Occurrence:

Tier 0 (No Data)	Essentially no data were available of this type for this class or subclass of TOrCs, including data that could be used for modeling.
Tier 1	For the majority of TOrCs in this class or subclass, available data are derived from single peer-reviewed occurrence studies.
Tier 2	For the majority of TOrCs in this class or subclass, available data are derived from multiple peer-reviewed occurrence studies employing appropriate analytical protocols such as isotope dilution mass spectrometry.
Tier 3	For the majority of TOrCs in this class or subclass, available data are derived from large, nationally-representative occurrence studies employing analytical protocols of the highest caliber.

10.5 Mobility of Biosolids-Borne Trace Organic Chemicals in Soils

Considerable data were available with respect to understanding the potential mobility of the targeted TOrCs in biosolids-amended soils. In particular, physicochemical parameters such as soil-water partition coefficients and octanol-water partition coefficients were available for many of the targeted TOrCs. However, some of the TOrCs examined in this study (i.e., tetracycline, ciprofloxacin, and perfluorochemicals) do not necessary follow the traditional hydrophobic organic contaminant partitioning paradigm. Clearly, *appropriate* applications of existing modeling approaches and alternative modeling approaches are needed to adequately describe the mobility behavior of TOrCs in biosolids-amended soils.

Understanding the mobility of biosolids-borne TOrCs in biosolids-amended soils requires working beyond the laboratory-scale, and evaluating mobility in bench-scale column studies and, especially, in field-scale experiments. Unfortunately, few such studies exist for several of the targeted TOrCs. Some studies indicated that some TOrCs can leach from fields, particularly when the applied biosolids are not dewatered, whereas other TOrCs (e.g., polybrominated diphenyl ethers, synthetic musks, and some steroidal chemicals) exhibited low leaching potential. More comprehensive bench and field-scale studies (with respect to analytes) are needed to accurately represent the real-world conditions under which biosolids are applied.

Bench and field-scale experiments on TOrC mobility would also help address another major data gap identified for nearly all of the TOrCs. The issue of irreversible sorption (chemisorption) and desorption of the targeted TOrCs from soils was not addressed in most of the published mobility studies. Biosolids present a unique matrix in soils, and for the same reason that many TOrCs may not fit the traditional partitioning paradigms (the presence of active functional groups), there is likely a greater potential for these TOrCs to become irreversibly bound to either soil organic matter or the biosolids-derived organic matter. Conversely, the binding of deprotonated TOrCs (many of the targeted TOrCs exist as anions at environmentally relevant pH values) can be substantially less than predicted from the hydrophobicity of the neutral form. In general, neither pH-dependent sorption nor the potential for irreversible (or strongly hysteretic) sorption are considered in traditional mobility models, many of which assume reversible sorption. Further studies of desorption in all compound classes are required to identify the TOrCs for which this is an important issue.

Table 10-4 provides a summary of the mobility data availability with respect to the high priority TOrCs included in this study. The decision criteria used to bin the TOrC classes or subclasses are provided, though in some cases, expert judgment was required to consider data that did not readily fit within the framework developed. As is clear from the table, while a substantial body of knowledge exists regarding the mobility of many TOrCs in soils, some significant data gaps are still evident for some compound classes.

Table 10-4. Summary of Mobility Data Availability for the High Priority TOrCs.

Chemical Class	Data Availability
BFRs	Tier 1
PFCs and PFC Precursors	Tier 2
PPCPs: Antimicrobials	Tier 2
PPCPs: Antibiotics	Tier 2
PPCPs: Synthetic Musks	Tier 2
PPCPs: Other	Tier 0
Plasticizers	Tier 2
Steroidal Chemicals	Tier 2
Surfactants	Tier 2

Data Availability Ranking Decision Criteria ~ Mobility:

Tier 0 (No Data)	Essentially no data were available of this type for this class or subclass of TOrCs, including data that could be used for modeling.
Tier 1	For the majority of TOrCs in this class or subclass, physicochemical parameters have been measured (i.e., K_{ow}) that would enable sorption/mobility predications to be made using appropriate models.
Tier 2	For the majority of TOrCs in this class or subclass, mobility has been evaluated in laboratory-based spiking studies employing actual soils or sediments and appropriate analytical protocols.
Tier 3	For the majority of TOrCs in this class or subclass, realistic and nationally-relevant field-scale studies been conducted evaluating the mobility (i.e., leaching) from biosolids-amended soils.

10.6 Persistence of Biosolids-Borne Trace Organic Chemicals in Soils

The persistence of biosolids-borne TOrCs in soils is a result of many processes, but biodegradation is generally considered the dominant process affecting TOrC attenuation in biosolids-amended soils. For most of the high priority TOrCs, no soil biodegradation data were identified from the peer-reviewed literature. In the absence of these data, it may be possible to make qualitative predictions of biodegradability for some TOrCs based on data from analogous chemicals. In particular, while certain PPCPs and steroidal chemicals have benefited from research in both aquatic and soil systems, others lack such data in environmental systems most relevant to biosolids amendment. This deficit was true for all BFRs as well as many of the PFC precursors of concern. When soil or biosolids-specific transformation data were unavailable, data from aquatic systems were considered as general indicators of recalcitrance, though their applicability to biosolids-amended soils is tenuous.

Environmental factors such as pH, moisture content, metal cations, temperature, and bacterial cell concentration all can affect biodegradation rates. The effects of such factors and the impact of different soil types or biosolids loading rates on attenuation need to be further investigated for most targeted TOrCs. Literature for antimicrobials and antibiotics indicate recalcitrance and slow biodegradation in soil systems and a dependence on site characteristics such as biosolids content, aerobic conditions, and soil depth. Biodegradation rates of steroidal chemicals can be favorably impacted by the presence of biosolids, increased temperatures, and adequate (but not excessive) water content in soils. Unfortunately, degradation data for many of the TOrCs included in this study are lacking for soil or biosolids-amended soils. Hence, discerning the current rate-limiting TOrCs that could mandate solids loading or biosolids application rates is difficult. Most TOrCs are transformed to less toxic intermediates, but aqueous studies of some of the targeted TOrCs highlight the potential for more toxic degradation products, particularly for the polybrominated diphenyl ethers and perfluorochemical precursors. Whether these processes also occur in soil systems remains unclear. Indeed, the behavior of likely degradation products of target compounds is little studied and deserves research attention.

Future TOrC biodegradation research should focus on soil and biosolids-amended soil systems to better understand the risks associated with biosolids-borne TOrCs in the environment. Data pertaining to BFRs, PFC precursors, plasticizers, and surfactants would benefit most from additional biosolids-focused research, but so would most of the TOrCs targeted in this analysis.

Table 10-5 provides a summary of the persistence data availability with respect to the high priority TOrCs included in this study. The decision criteria used to bin the TOrC classes or subclasses are provided, though in some cases, expert judgment was required to consider data that did not readily fit within the framework developed. As is clear from the table, significant data gaps with respect to persistence are evident for many of the TOrCs included in this study.

Table 10-5. Summary of Persistence Data Availability for the High Priority TOrCs.

Chemical Class	Data Availability
BFRs	Tier 1
PFCs and PFC Precursors	Tier 1
PPCPs: Antimicrobiall Agents	Tier 3
PPCPs: Antibiotics	Tier 1
PPCPs: Synthetic Musks	Tier 3
PPCPs: Other	Tier 0
Plasticizers	Tier 1
Steroidal Chemicals	Tier 2
Surfactants	Tier 0

Data Availability Ranking Decision Criteria ~ Persistence:

Tier 0 (No Data)	Essentially no data were available of this type for this class or subclass of TOrCs, including data that could be used for modeling.
Tier 1	For the majority of TOrCs in this class or subclass, biodegradation studies have been conducted (disregarding of incubation medium or environment).
Tier 2	For the majority of TOrCs in this class or subclass, biodegradation studies have been conducted in soils.
Tier 3	For the majority of TOrCs in this class or subclass, field-scale persistence studies have been conducted.

10.7 Bioaccumulation of Biosolids-Borne Trace Organic Chemicals in Soils

Data on the bioaccumulation of biosolids-borne TOrCs in plants and animals were examined, but few useful data sets were found. Bioaccumulation of some of the TOrCs has been documented, but few studies examined bioaccumulation and bioavailability specifically in biosolids-amended soils. Since data derived from biosolids-amended systems were extremely limited, general accumulation data from soils was also evaluated. Bioaccumulation data from sediments (particularly for animals) were also compiled, but the relevance of these studies to biosolids-amended soils is questionable.

Some of the targeted TOrCs (tetracycline antibiotics, antimicrobials, fluoroquinolones, and synthetic musks, brominated flame retardants) can accumulate in a variety of plants including grass, green onions, cabbage, corn, alfalfa, lettuce, radish, zucchini, and carrots. Data for other compound classes were generally absent. More data were available for the bioaccumulation and bioavailability of TOrCs in animals, particularly invertebrates such as earthworms. Unfortunately, many of the studies identified did not provide significant detail as to the exposure conditions, making the modeling of the bioaccumulation highly problematic. Parameters such as BAFs and biota-soil accumulation factors (BSAFs) are meant to facilitate comparisons of bioavailability between chemicals and between sites, but, factor units were not always provided or were inconsistent, making comparisons difficult.

Some TOrCs such as PFCs do not accumulate in lipids, rendering lipid normalization inappropriate. In addition, the affinity of other TOrCs for the solid phase (i.e., tetracyclines) does not necessarily depend on organic carbon, rendering organic carbon normalization problematic. These factors point to a need for consistency in measuring and reporting data to facilitate comparisons among TOrCs. Factors that should be considered include the organisms used (i.e., standard organisms), how the chemical is introduced to the organism, use of environmentally relevant conditions, and standardization of units and methods of normalization to calculate BAF and BSAF values.

The bioaccumulation data evaluated in this study focused on the uptake of TOrCs from biosolids amended soils to either plants or animals. While *biomagnification*, a process by which the body burden of the TOrC increases as the trophic level increases, of TOrCs has not been demonstrated from biosolids-amended soil, it may occur for some TOrCs depending on their chemical properties and the food chain pathway. Consideration of such processes may be important for risk models. For example, a recent study suggested the most sensitive pathway for trophic transfer of triclocarban was from earthworms to birds (Snyder, 2009). However, this assessment was based on model predictions and thus would need to be verified by laboratory or field experiments, and such data are extremely limited.

Table 10-6 provides a summary of the bioaccumulation data availability with respect to the high priority TOrCs included in this study. The decision used to bin the TOrC classes or subclasses are provided, though in some cases, expert judgment was required to consider data that did not readily fit within the framework developed. As is clear from the table, significant data gaps with respect to bioaccumulation are evident for many of the TOrCs included in this study.

Table 10-6. Summary of Bioaccumulation Data Availability for the High Priority TOrCs.

Chemical Class	Data Availability
BFRs	Tier 2
PFCs and PFC Precursors	Tier 0
PPCPs: Antimicrobials	Tier 1
PPCPs: Antibiotics	Tier 0
PPCPs: Synthetic Musks	Tier 2
PPCPs: Other	Tier 0
Plasticizers	Tier 1
Steroidal Chemicals	Tier 1
Surfactants	Tier 1

Data Availability Ranking Decision Criteria ~ Bioaccumulation:

Tier 0 (No Data)	Essentially no data were available of this type for this class or subclass of TOrCs, including data that could be used for modeling.
Tier 1	For the majority of TOrCs in this class or subclass, physicochemical parameters have been measured (i.e., K_{ow}) that would enable bioaccumulation potential to be assessed in both plants and animals.
Tier 2	For the majority of TOrCs in this class or subclass, bioaccumulation studies have been conducted for either plants or animals in spiked soil or sediment systems using appropriate species and analytical protocols.
Tier 3	For the majority of TOrCs in this class or subclass, realistic field-scale monitoring studies have been conducted evaluating bioaccumulation in both plants and animals from biosolids-amended soils.

10.8 Toxicity of Biosolids-Borne Trace Organic Chemicals in Soils

Even though the focus of this study was on the terrestrial environment, both human and ecological toxicity data were sought for the high priority TOrCs. Relevant human toxicity values were identified for less than half of the targeted TOrCs, though an exhaustive review may have identified more human toxicity data. Furthermore, the data gathered should be further scrutinized with regard to the confidence they engender. For example, substantial bodies of data and expert scientific review were involved in the development of toxicity values for PFCs and bisphenol A, but little chemical-specific information was available for development of the 4-cumylphenol threshold of toxicological concern. For some TOrCs the mode of action, particularly if additive toxicity is possible, should be evaluated in more detail.

Ecotoxicological data for the targeted TOrCs were sought, but relevant soil and sediment toxicity data were found for only a few of the high priority TOrCs. Even when relevant ecotoxicity studies were identified, they were limited in terms of quantity, study quality, toxicological endpoints investigated, and number of species and taxa evaluated. A significant proportion of the available studies were conducted in sediment, and the applicability of these studies for soils, much less biosolids-amended soils, is highly questionable. Significant data gaps exist with respect to the toxicity of the targeted TOrCs in terrestrial environments, particularly in biosolids-amended soils. Comparatively, there are substantial volumes of toxicity data for many of the targeted TOrCs in aquatic environments, and WERF-sponsored evaluations of these data are currently underway. However, while some of the exposure pathways relevant for biosolids-amended soils are aquatic in nature, aquatic toxicity data were not sought in this review (with the noted exception of sediment studies). Even including the aquatic toxicity data, the known and potential modes of action of the targeted TOrCs in ecological receptors is far from comprehensive and should be expanded in future efforts. Clearly, additional studies are needed to examine the terrestrial toxicity of the targeted TOrCs in biosolids-amended soils, particularly studies that include trophic transfer and toxicity to higher trophic level organisms.

Table 10-7 provides a summary of the toxicity data availability with respect to the high priority TOrCs included in this study. The decision criteria used to bin the TOrC classes or subclasses is indicated, though in some cases, expert judgment was required to consider data that did not readily fit within the framework developed. Significant data gaps with respect to toxicity, particularly ecological toxicity, are evident for many of the TOrCs included in this study.

Table 10-7. Summary of Toxicity Data Availability for the High Priority TOrCs.

Chemical Class	Toxicity Data Availability	
	Human	Ecological
BFRs	Tier 0	Tier 0
PFCs and PFC Precursors	Tier 0	Tier 0
PPCPs: Antimicrobials	Tier 0	Tier 0
PPCPs: Antibiotics	Tier 2[1]	Tier 0
PPCPs: Synthetic Musks	Tier 1	Tier 0
PPCPs: Other	Tier 2[1]	Tier 0
Plasticizers	Tier 1	Tier 0
Steroidal Chemicals	Tier 2[1]	Tier 0
Surfactants	Tier 1	Tier 0

[1] The potential health effects of therapeutic use of drugs may be considered to be well-characterized due to testing required during drug development and registration. However, potential health effects of non-therapeutic exposure (e.g., longer term exposures, sub-therapeutic doses, unusual routes of exposure) and exposure of non-target populations (e.g., people for whom the use of the drugs is contraindicated) are often unknown.

Data Availability Ranking Decision Criteria ~ Human Toxicity:

Tier 0 (No Data)	Essentially no data were available of this type for this class or subclass of TOrCs, including data that could be used for modeling.
Tier 1	For the majority of TOrCs in this class or subclass, substantial animal toxicological data (both acute and chronic) are available.
Tier 2	For the majority of TOrCs in this class or subclass, substantial human toxicological data are available.
Tier 3	For the majority of TOrCs in this class or subclass, human health benchmark (HHB) values been derived. Note: In some cases, HHBs are derived simply by applying safety/uncertainty factors to the therapeutic dose. Thus, the existence of an HHB does not necessarily imply the availability of a substantial body of toxicological information useful for risk assessment of soil exposure for the general population.

Data Availability Ranking Decision Criteria ~ Ecological Toxicity:

Tier 0 (No Data)	Essentially no data were available of this type for this class or subclass of TOrCs, including data that could be used for modeling.
Tier 1	For the majority of TOrCs in this class or subclass, relevant soil or sediment ecotoxicological data representing multiple taxa are available in either sediment or soil.
Tier 2	For the majority of TOrCs in this class or subclass, relevant ecotoxicological data representing multiple taxa are available in soil systems.
Tier 3	For the majority of TOrCs in this class or subclass, ecotoxicological data are available for these TOrCs in biosolids-amended soils

10.9 Impacts on Soil Microbial Communities

Though biosolids are most often land applied in agricultural settings where land management practices are expected to alter natural microbial ecosystems, some TOrCs may have toxicological effects on soil macrobiota or soil microbial communities over and above the effects of the biosolids. Microbial impacts can be measured by examining alterations in community composition, metabolic function, and diversity. Data pertaining to the microbial impacts, much less microbial impacts in biosolids-amended soils, of many of the targeted TOrCs were not available. Where possible, data for analogous chemicals within the identified classes were evaluated. Studies identified for the targeted TOrCs demonstrate a variety of effects on soil microbial structure and function, though data derived from biosolids-amended soils were limited. The types of effects observed include suppression of soil nitrification rates, increases in antibiotic resistance, and other general changes to community structure, metabolism, and diversity. However, few generalizations can be made, even within specific classes of TOrCs. Exposure of microbial communities to some pharmaceuticals and personal care product ingredients increased microbial biomass, richness, and the sizes of certain bacterial populations, whereas others increased the presence of antibiotic resistance genes (ARGs).

A better understanding of potential selective pressures resulting from the introduction of the TOrCs in a biosolids matrix is needed to adequately address the risk and acceptable loads associated with the land application of biosolids. This includes consortia structural properties, propagation of ARGs, and functional processes such as attenuation and nutrient cycling. Few data are available for many of the targeted TOrCs, suggesting a significant need for research into the potential effects of the targeted TOrC on microbial systems in managed soils.

Table 10-8 provides a summary of the microbial impact data availability with respect to the high priority TOrCs included in this study. The decision criteria used to bin the TOrC classes or subclasses are provided, though in some cases, expert judgment was required to consider data that did not readily fit within the framework developed. As is clear from the table, significant data gaps with respect to microbial impacts are evident for many of the TOrCs included in this study.

Table 10-8. Summary of Microbial Impacts Data Availability for the High Priority TOrCs.

Chemical Class	Data Availability
BFRs	Tier 0
PFCs and PFC Precursors	Tier 0
PPCPs: Antimicrobials	Tier 1
PPCPs: Antibiotics	Tier 1
PPCPs: Synthetic Musks	Tier 0
PPCPs: Other	Tier 0
Plasticizers	Tier 0
Steroidal Chemicals	Tier 0
Surfactants	Tier 0

Data Availability Ranking Decision Criteria ~ Microbial Impacts:

Tier 0 (No Data)	Essentially no data were available of this type for this class or subclass of TOrCs, including data that could be used for modeling.
Tier 1	For the majority of TOrCs in this class or subclass, ecotoxicological data were available for microorganisms.
Tier 2	For the majority of TOrCs in this class or subclass, ecotoxicological data were available for microorganisms in soil systems.
Tier 3	For the majority of TOrCs in this class or subclass, ecotoxicological data were available for microorganisms in biosolids-amended soils.

10.10 Overall Data Gaps

Substantial data gaps exist for many processes important to understanding the risk of biosolids-borne TOrCs in soil environments. Some data are available for particular processes (e.g., sorption), but few data were found specifically with respect to biosolids-amended soils. Very few studies were identified that were intentionally designed to address the fate, transport, bioaccumulation, and toxicity of TOrCs in biosolids-amended soils under well-controlled conditions. The complexity of the biosolids matrix is often ignored in many studies, and the potential for irreversibly bound residues of TOrCs in biosolids-amended soils has not been well-characterized, or appropriately modeled. Bench-scale column studies may be appropriate avenues of research for addressing specific questions related to the fate and transport of biosolids-borne TOrCs in soils. *The most significant data gap, however, is the absence of human toxicological and ecotoxicological data as well as biotransfer data for ecological receptors. Well-studied long-term field plots to which biosolids have been applied would also help in exposure and effects evaluations, as well as validation of the risk model predictions.* Such field studies can incorporate the complexities often ignored in more controlled laboratory settings, and offer the possibility of conducting studies with multiple objectives (i.e., mobility, bioaccumulation, and volatilization) under identical conditions. Additional data are needed to reduce the uncertainty in assessing the risk of biosolids-borne TOrCs in the terrestrial environment.

REFERENCES

3M (2000). Soil adsorption/desorption study of potassium perfluorooctane sulfonate (PFOS). St. Paul, MN, 3M. United States Environmental Protection Agency Docket AR 226-1030a 030.

Agyin-Birikorang, S., M. Miller, and G.A. O'Connor (submitted). Retention-release characteristics of triclocarban and triclosan in biosolids, soils, and biosolids-amended soils.

Ahel, M. and W. Giger (1993). Partitioning of alkylphenols and alkylphenol polyethoxylates between water and organic-solvents. *Chemosphere.* **26**(8): 1471-1478.

Al-Rajab, A., L. Sabourin, A. Scott, D. Lapen, and E. Topp (2009). Impact of biosolids on the persistence and dissipation pathways of triclosan and triclocarban in an agricultural soil. *Science of the Total Environment.* **407**(23): 5978-85.

Albiach, R., R. Canet, F. Pomares, and F. Ingelmo (2000). Microbial biomass content and enzymatic activities after the application of organic amendments to a horticultural soil. *Bioresource Technology.* **75**(1): 43-48.

Andersen, J. (2001). Disposal and recycling routes for sewage sludge. Part 3 - Scientific and Technical Subcomponent report. R. f. t. E. Commission.

Aranami, K. and J.W. Readman (2007). Photolytic degradation of triclosan in freshwater and seawater. *Chemosphere.* **66**(6): 1052-1056.

Arbeli, Z. and Z. Ronen (2003). Enrichment of a microbial culture capable of reductive debromination of the flame retardant tetrabromobisphenol-A, and identification of the intermediate metabolites produced in the process. *Biodegradation.* **14**(6): 385-95.

Arnold, R., S. Teske, M. Tomanek, J. Engstrom, C. Leung, J. Zhang, Q. Banihani, D. Quanrud, W. Ela, and A. Sáez (2008). Fate of polybrominated diphenyl ethers during wastewater treatment/polishing and sludge stabilization/disposal. *Annals of teh New York Academy of Science.* **1140**: 394-411.

Arnot, J.A. and F. Gobas (2006). A review of bioconcentration factor (BCF) and bioaccumulation factor (BAF) assessments for organic chemicals in aquatic organisms. *Environmental Reviews.* **14**(4): 257-297.

Arp, H.P.H., C. Niederer, and K.U. Goss (2006). Predicting the partitioning behavior of various highly fluorinated compounds. *Environmental Science and Technology.* **40**(23): 7298-7304.

Artola-Garicano, E., T.L. Sinnige, I. van Holsteijn, W.H. Vaes, and J.L. Hermens (2003). Bioconcentration and acute toxicity of polycyclic musks in two benthic organisms *(Chironomus riparius* and *Lumbriculus variegatus). Environmental Toxicology and Chemistry.* **22**(5): 1086-1092.

ATSDR. (2009a). Minimal Risk Levels (MRLs). Department of Health and Human Services, Public Health Service, Agency for Toxic Substances and Disease Registry (ATSDR). Atlanta, Georgia. from http://www.atsdr.cdc.gov/mrls/.

ATSDR. (2009b). Toxicological Profiles. Department of Health and Human Services, Public Health Service, Agency for Toxic Substances and Disease Registry (ATSDR). Atlanta, Georgia. from http://www.atsdr.cdc.gov/toxpro2.html.

Auerbach, E.A., E.E. Seyfried, and K.D. McMahon (2007). Tetracycline resistance genes in activated sludge wastewater treatment plants. *Water Research.* **41**(5): 1143-51.

Ayscough, N.J., J. Fawell, G. Franklin, and W. Young (2000). Review of human pharmaceuticals in the environment. Bristol, Environment Agency. P390.

Baird, B.H. and D.C. White (1985). Biomass and community structure of the abyssal microbiota determined from the ester-linked phospholipids recovered from Venezuela basin and Puerto Rico trench sediments. *Marine Geology.* **68**(1-4): 217-231.

Balk, F. and R.A. Ford (1999a). Environmental risk assessment for the polycyclic musks AHTN and HHCB in the EU - I. Fate and exposure assessment. *Toxicology Letters.* **111**(1-2): 57-79.

Balk, F. and R.A. Ford (1999b). Environmental risk assessment for the polycyclic musks AHTN and HHCB in the EU - II. Effect assessment and risk characterisation. *Toxicology Letters.* **111**(1-2): 81-94.

Banks, M., A. Schwab, N. Cofield, J. Alleman, M. Switzenbaum, J. Shalabi, and P. Williams (2006). Biosolids-amended soils: Part I. Effect of biosolids application on soil quality and ecotoxicity. *Water Environment Research.* **78**(11): 2217-30.

Bao, Y.Y., Q.X. Zhou and Y.Y. Wang (2009). Adsorption characteristics of tetracycline by two soils: assessing role of soil organic matter. *Australian Journal of Soil Research.* **47**(3): 286-295.

Barnabe, S., I. Beauchesne, D. Cooper, and J. Nicell (2008). Plasticizers and their degradation products in the process streams of a large urban physicochemical sewage treatment plant. *Water Research.* **42**(1-2): 153-162.

Barron, L., J. Havel, M. Purcell, M. Szpak, B. Kelleher, and B. Paull (2009). Predicting sorption of pharmaceuticals and personal care products onto soil and digested sludge using artificial neural networks. *Analyst.* **134**(4): 663-670.

Berset, J.D., P. Bigler, and D. Herren (2000). Analysis of nitro musk compounds and their amino metabolites in liquid sewage sludges using NMR and mass spectrometry. *Analytical Chemistry.* **72**(9): 2124-2131.

Biggers, W.J. and H. Laufer (2004). Identification of juvenile hormone-active alkylphenols in the lobster *Homarus americanus* and in marine sediments. *Biological Bulletin.* **206**(1): 13-24.

Bolz, U., H. Hagenmaier, and W. Korner (2001). Phenolic xenoestrogens in surface water, sediments, and sewage sludge from Baden-Wurttemberg, south-west Germany. *Environmental Pollution.* **115**(2): 291-301.

Bonin, J.L. and M.J. Simpson (2007). Sorption of steroid estrogens to soil and soil constituents in single- and multi-sorbate systems. *Environmental Toxicology and Chemistry.* **26**(12): 2604-2610.

Bossi, R., J. Strand, O. Sortkjaer and M.M. Larsen (2008). Perfluoroalkyl compounds in Danish wastewater treatment plants and aquatic environments. *Environment International.* **34**(4): 443-450.

Boxall, A., L. Fogg, P. Blackwell, P. Kay, and E. Pemberton (2002). Review of Veterinary Medicines in the Environment. Bristol, United Kingdom, Environment Agency of England and Wales. R&D Technical Report P6-012/8/TR.

Boxall, A.B.A., L.A. Fogg, P. Kay, P.A. Blackwell, E.J. Pemberton, and A. Croxford (2003). Prioritisation of veterinary medicines in the UK environment. *Toxicology Letters.* **142**(3): 207-218.

Boxall, A.B.A., P. Johnson, E.J. Smith, C.J. Sinclair, E. Stutt, and L.S. Levy (2006). Uptake of veterinary medicines from soils into plants. *Joranal of Agricultural and Food Chemistry.* **54**(6): 2288-2297.

Bradley, P.M., L.B. Barber, F.H. Chapelle, J.L. Gray, D.W. Kolpin, and P.B. McMahon (2009). Biodegradation of 17Î²-Estradiol, Estrone and Testosterone in Stream Sediments. *Environmental Science and Technology.* **43**(6): 1902-1910.

Brooks, J., C. Gerba, and I. Pepper (2007a). Diversity of aerosolized bacteria during land application of biosolids. *Journal of Applied Microbiology.* **103**(5): 1779-90.

Brooks, J., S. Maxwell, C. Rensing, C. Gerba, and I. Pepper (2007b). Occurrence of antibiotic-resistant bacteria and endotoxin associated with the land application of biosolids. *Canadian Journal of Microbiology.* **53**(5): 616-22.

Brown, S., D. Devin-Clarke, M. Doubrava, and G. O'Connor (2009). Fate of 4-nonylphenol in a biosolids amended soil. *Chemosphere.* **75**(4): 549-554.

Burns, D.C., D.A. Ellis, H. Li, C.J. McMurdo, and E. Webster (2008). Experimental pK(a) Determination for Perfluorooctanoic Acid (PFOA) and the Potential Impact of pK(a) Concentration Dependence on Laboratory-Measured Partitioning Phenomena and Environmental Modeling. *Environmental Science and Technology.* **42**(24): 9283-9288.

Cal/Ecotox. (2009). The California Wildlife Biology, Exposure Factor, and Toxicity Database (Cal/Ecotox). California Office of Environmental Health Hazard Assessment (OEHHA), Sacramento, CA. from http://www.oehha.ca.gov/scripts/cal_ecotox/chemicaldescription.asp.

Canesi, L., L.C. Lorusso, C. Ciacci, M. Betti, and G. Gallo (2005). Effects of the brominated flame retardant tetrabromobisphenol-A (TBBPA) on cell signaling and function of Mytilus hemocytes: Involvement of MAP kinases and protein kinase C. *Aquatic Toxicology.* **75**(3): 277-287.

Cantero, M., S. Rubio, and D. Perez-Bendito (2004). Determination of non-ionic polyethoxylated surfactants in sewage sludge by coacervative extraction and ion trap liquid chromatography-mass spectrometry. *Journal of Chromatography A.* **1046**(1-2): 147-153.

Capdevielle, M., R. Van Egmond, M. Whelan, D. Versteeg, M. Hofmann-Kamensky, J. Inauen, V. Cunningham, and D. Woltering (2008). Consideration of exposure and species sensitivity of triclosan in the freshwater environment. *Integrated Environmental Assessment and Management.* **4**(1): 15-23.

Carballa, M., G. Fink, F. Omil, J.M. Lema, and T. Ternes (2008). Determination of the solid-water distribution coefficient (Kd) for pharmaceuticals, estrogens and musk fragrances in digested sludge. *Water Research.* **42**(1-2): 287-295.

Carballa, M., G. Manterola, L. Larrea, T. Ternes, F. Omil, and J.M. Lema (2007a). Influence of ozone pre-treatment on sludge anaerobic digestion: Removal of pharmaceutical and personal care products. *Chemosphere.* **67**(7): 1444-1452.

Carballa, M., F. Omil, T. Ternes, and J.M. Lema (2007b). Fate of pharmaceutical and personal care products (PPCPs) during anaerobic digestion of sewage sludge. *Water Research.* **41**(10): 2139-2150.

Carmosini, N. and L.S. Lee (2008a). Partitioning of fluorotelomer alcohols to octanol and different sources of dissolved organic carbon. *Environmental Science and Technology.* **42**(17): 6559-65.

Carmosini, N. and L.S. Lee (2008b). Sorption and Degradation of Selected Pharmaceuticals in Soil and Manure.*In* Fate of Pharmaceuticals in the Environmenta and in Water Treatment Systems. D. S. Aga. Boca Raton, FL, Taylor & Francis Group.

Carmosini, N. and L.S. Lee (2009). Ciprofloxacin sorption by dissolved organic carbon from reference and bio-waste materials. *Chemosphere.* **77**: 813-820.

Carrasquillo, A.J., G.L. Bruland, A.A. Mackay, and D. Vasudevan (2008). Sorption of Ciprofloxacin and Oxytetracycline Zwitterions to Soils and Soil Minerals: Influence of Compound Structure. *Environmental Science and Technology.* **42**(20): 7634-7642.

Cartwright, C.D., S.A. Owen, I.P. Thompson, and R.G. Burns (2000). Biodegradation of diethyl phthalate in soil by a novel pathway. *FEMS Microbiology Letters.* **186**(1): 27-34.

CEC (1991a). Council Directive 91/692/EEC of 23 December 1991 standardizing and rationalizing reports on the implementation of certain directives relating to the environment (plus various amendments implementing the Directive). *Official Journal of the European Communities.* **L230**.

CEC (1991b). Council Directive concerning the placing of plant protection products on the market (91/414 EEC) (plus various amendments implementing the Directive). *Official Journal of the European Communities.* **L230**.

CEC (1996). Technical Guidance Document in support of the Commission Directive 93/67/EEC on risk assessment or new notified substances and Commission Regulation EEC # 1488/94 on risk assessment for existing substances. Ispra, Italy. E. C. Bureau.

CEC (2000). CSTEE opinion on the available scientific approaches to assess the potential effects and risk of chemicals on terrestrial ecosystems. D. o. G. H. a. C. P. European Commission, Scientific Committee on Toxicity, Ecotoxicity and the Environment.

CEC (2003). Second Edition of the Technical Guidance Document on Risk Assessment of Chemical Substances following European Regulation and Directives. JRC-ISPRA, Italy. E. C. Bureau.

Cha, J.M. and A.M. Cupples (2009). Detection of the antimicrobials triclocarban and triclosan in agricultural soils following land application of municipal biosolids. *Water Research.* **43**(9): 2522-2530.

Chau, Y.K., S.Z. Zhang, and R.J. Maguire (1992). Occurrence of butyltin species in sewage and sludge in Canada. *Science of the Total Environment.* **121**: 271-281.

Chee-Sanford, J., R. Mackie, S. Koike, I. Krapac, Y. Lin, A. Yannarell, S. Maxwell, and R. Aminov (2009). Fate and transport of antibiotic residues and antibiotic resistance genes following land application of manure waste. *Journal of Environmental Quality.* **38**(3): 1086-108.

Chee-Sanford, J.C., R.I. Aminov, I.J. Krapac, N. Garrigues-JeanJean, and R.I. Mackie (2001). Occurrence and diversity of tetracycline resistance genes in lagoons and groundwater underlying two swine production facilities. *Applied Environmental Microbiology.* **67**(4): 1494–1502.

Chen, G., B.A. Vanegas, and K.A. Strevett (2001). Naphthalene, phenanthrene and surfactant biodegradation. *Biodegradation.* **12**(6): 433-442.

Ciparis, S. and R.C. Hale (2005). Bioavailability of polybrominated diphenyl ether flame retardants in biosolids and spiked sediment to the aquatic oligochaete, *Lumbriculus variegatus*. *Environmental Toxicology and Chemistry.* **24**(4): 916-925.

Colucci, M., H. Bork, and E. Topp (2001). Persistence of estrogenic hormones in agricultural soils: I. 17Beta-estradiol and estrone. *Journal of Environmental Quality.* **30**(6): 2070-6.

Colucci, M. and E. Topp (2001). Persistence of estrogenic hormones in agricultural soils: II. 17Alpha-ethynylestradiol. *Journal of Environmental Quality.* **30**(6): 2077-80.

Connell, D.W. and R.D. Markwell (1990). Bioaccumulation in the soil to earthworm system. *Chemosphere.* **20**(1-2): 91-100.

Coogan, M.A. and T.W. La Point (2008). Snail bioaccumulation of triclocarban, triclosan, and methyltriclosan in a North Texas, USA, stream affected by wastewater treatment plant runoff. *Environmental Toxicology and Chemistry.* **27**(8): 1788-1793.

Cordova-Kreylos, A.L. and K.M. Scow (2007). Effects of ciprofloxacin on salt marsh sediment microbial communities. *Isme J.* **1**(7): 585-95.

Czajka, C.P. and K.L. Londry (2006). Anaerobic biotransformation of estrogens. *Science of the Total Environment.* **367**(2-3): 932-941.

D'Ascenzo, G., A. Di Corcia, A. Gentili, R. Mancini, R. Mastropasqua, M. Nazzari, and R. Samperi (2003). Fate of natural estrogen conjugates in municipal sewage transport and treatment facilities. *Science of the Total Environment.* **302**(1-3): 199-209.

D'Eon, J.C., P.W. Crozier, V.I. Furdui, E.J. Reiner, E.L. Libelo, and S.A. Mabury (2009). Observation of a Commercial Fluorinated Material, the Polyfluoroalkyl Phosphoric Acid Diesters, in Human Sera, Wastewater Treatment Plant Sludge, and Paper Fibers. *Environmental Science and Technology.* **43**(12): 4589-4594.

Daughton, C.G. and T.A. Ternes (1999). Pharmaceuticals and personal care products in the environment: agents of subtle change? *Environmental Health Perspectives.* **107 Suppl 6**: 907-38. 938daughton/daughton-full.html 938daughton/abstract.html.

de Weert, J., M. Streminska, D. Hua, T. Grotenhuis, A. Langenhoff, and H. Rijnaarts (2010). Nonylphenol mass transfer from field-aged sediments and subsequent biodegradation in reactors mimicking different river conditions. *Journal of Soils and Sediments.* **10**: 77-88.

de Wit, C.A. (2002). An overview of brominated flame retardants in the environment. *Chemosphere.* **46**(5): 583-624.

den Hollander, H., J.C.H. van Eijkeren, and D. van de Meent (2004). SimpleBox 3.0: A nested multimedia fate model for evaluating the environmental fate of chemicals. Bilthoven, The Netherlands, Nation Institute of Public Health and the Environment. RIVM report no 601200003. .

Difrancesco, A.M., P.C. Chiu, L.J. Standley, H.E. Allen, and D.T. Salvito (2004). Dissipation of fragrance materials in sludge-amended soils. *Environmental Science and Technology.* **38**(1): 194-201.

Dinglasan, M.J., Y. Ye, E.A. Edwards, and S.A. Mabury (2004). Fluorotelomer alcohol biodegradation yields poly- and perfluorinated acids. *Environmental Science and Technology.* **38**(10): 2857-64.

Ditoro, D.M., C.S. Zarba, D.J. Hansen, W.J. Berry, R.C. Swartz, C.E. Cowan, S.P. Pavlou, H.E. Allen, N.A. Thomas, and P.R. Paquin (1991). Technical basis for establishing sediment quality criteria for nonionic organic-chemicals using equilibrium partitioning. *Environmental Toxicology and Chemistry.* **10**(12): 1541-1583.

DOE. (2009). Risk Assessment Information System. United States Department of Energy (DOE), Office of Environmental Management, Oak Ridge Operations (ORO) Office, through a contract with Bechtel Jacobs Company LLC. Chemical-specific values current as of May 2007. from http://rais.ornl.gov/.

Domene, X., S. Chelinho, and J.P. Sousa (In press). Effects of nonylphenol on a soil community using microcosms. *Journal of Soils and Sediments.*

Drakopoulos, A.I. and P.C. Ioannou (1997). Spectrofluorimetric study of the acid-base equilibria and complexation behavior of the fluoroquinolone antibiotics ofloxacin, norfloxacin, ciprofloxacin and pefloxacin in aqueous solution. *Analytica Chimica Acta.* **354**(1-3): 197-204.

Drewes, J.E., E. Dickenson, and S.A. Snyder (2008). *Contributions of Household Chemicals to Sewage and their Relevance to Municipal Wastewater Systems and the Environment.* 03-CTS-21U. Water Environment Research Foundation, Alexandria, VA.

Drillia, P., K. Stamatelatou, and G. Lyberatos (2005). Fate and mobility of pharmaceuticals in solid matrices. *Chemosphere.* **60**(8): 1034-1044.

Duft, M., C. Schmitt, J. Bachmann, C. Brandelik, U. Schulte-Oehlmann, and J. Oehlmann (2007). Prosobranch snails as test organisms for the assessment of endocrine active chemicals - an overview and a guideline proposal for a reproduction test with the freshwater mudsnail Potamopyrgus antipodarum. *Ecotoxicology.* **16**(1): 169-182.

EC-BKH (2000). Towards the establishment of a priority list of substances for further evaluation of their role in endocrine disruption: preparation of a candidate list of substances as a basis for priority setting. Final report (incorporating corrigenda to final report dated 21 June 2000). Delft, The Netherlands. M0355008/1786Q/10/11/00.

EDSP (2005). Preliminary Draft Under Development - EDSP and Other Validation Chemicals at a Glance. (December 2005). Version 1.

Edwards, M., E. Topp, C.D. Metcalfe, H. Li, N. Gottschall, P. Bolton, W. Curnoe, M. Payne, A. Beck, S. Kleywegt, and D.R. Lapen (2009). Pharmaceutical and personal care products in tile drainage following surface spreading and injection of dewatered municipal biosolids to an agricultural field. *Science of the Total Environment.* **407**(14): 4220-4230.

Eisler, R. (2000a). Contaminant hazard reviews 1-35. Patuxent Wildlife Research Center, United States Geological Survey. Laurel, Maryland. from http://www.pwrc.usgs.gov/infobase/eisler/.

Eisler, R. (2000b). Handbook of chemical risk assessment: health hazards to humans, plants, and animals. Volume 2. Organics, Lewis Publishers. New York, NY.

Ejarrat, E., G. Marsh, A. Labandeira, and D. Barcelo (2008). Effect of sewage sludges contaminated with polybrominated diphenylethers on agricultural soils. *Chemosphere.* **71**(6): 1079-1086.

EPHC (2008). Australian Guidelines for Water Recycling: Managing Health and Environmental Risks (Phase 2): Augmentation of Drinking Water Supplies. . Adelaide, South Australia, Australian Health Ministers' Conference, Natural Resource Management Ministerial Council, Environment Protection and Heritage Council (EPHC).

European Commission SCF (2002). Opinion of the Scientific Committee on Food on Bisphenol A (Expressed on 17 April 2002). SCF/CS/PM/3936 Final 3 May 2002. European Commission Scientific Committee on Food (SCF).

Evandri, M.G., L.G. Costa, and P. Bolle (2003). Evaluation of brominated diphenyl ether-99 toxicity with *Raphidocelis subcapitata* and *Daphnia magna. Environmental Toxicology and Chemistry.* **22**(9): 2167-2172.

Fairbanks, B.C., G.A. O'Connor, and S.E. Smith (1985). Fate of di-2-(ethylhexyl)phthalate in three sludge-amended New Mexico soils. *Journal of Environmental Quality.* **14**(4): 479-482.

Federle, T.W., S.K. Kaiser, and B.A. Nuck (2002). Fate and effects of triclosan in activated sludge. *Environmental Toxicology and Chemistry.* **21**(7): 1330-1337.

Fendinger, N.J., D.C. Mcavoy, W.C. Eckhoff, and B.B. Price (1997). Environmental occurrence of polydimethylsiloxane. *Environmental Science and Technology.* **31**(5): 1555-1563.

Fent, K. (1989). Organotin Speciation in Municipal Waste-Water and Sewage-Sludge - Ecotoxiocological Consequences, Elsevier Sci Ltd.

Fent, K. (1996). Organotin compounds in municipal wastewater and sewage sludge: Contamination, fate in treatment process and ecotoxicological consequences, Elsevier Science.

Fent, K. and M.D. Muller (1991). Occurrence of organotins in municipal waste-water and sewage-sludge and behavior in a treatment-plant. *Environmental Science and Technology.* **25**(3): 489-493.

Figueroa, R.A., A. Leonard, and A.A. Mackay (2004). Modeling tetracycline antibiotic sorption to clays. *Environmental Science and Technology.* **38**(2): 476-483.

Foran, C.M., E.R. Bennett, and W.H. Benson (2000). Developmental evaluation of a potential non-steroidal estrogen: triclosan. *Marine Environmental Research.* **50**(1-5): 153-156.

Fujii, K., S. Kikuchi, M. Satomi, N. Ushio-Sata, and N. Morita (2002). Degradation of 17β-Estradiol by a Gram-Negative Bacterium Isolated from Activated Sludge in a Sewage Treatment Plant in Tokyo, Japan. *Applied and Environmental Microbiology.* **68**(4): 2057-60.

Garland, J.L. and A.L. Mills (1991). Classification and characterization of heterotrophic microbial communities on the basis of patterns of community-level sole-carbon-source utilization. *Applied and Environmental Microbiology.* **57**(8): 2351-2359.

Gerecke, A.C., W. Giger, P.C. Hartmann, N.V. Heeb, H.P.E. Kohler, P. Schmid, M. Zennegg, and M. Kohler (2006). Anaerobic degradation of brominated flame retardants in sewage sludge, Pergamon-Elsevier Science Ltd.

Gielen, G. (2007). The fate and effects of sewage-derived pharmaceuticals in soil. Christchurch, New Zealand, University of Canterbury. Doctor of Philosophy.

Gledhill, W.E. (1975). Biodegradation of 3,4,4'-Trichlorocarbanilide, TCC, in Sewage and Activated-Sludge. *Water Research.* **9**(7): 649-654.

Golet, E.M., A. Strehler, A.C. Alder, and W. Giger (2002). Determination of fluoroquinolone antibacterial agents in sewage sludge and sludge-treated soil using accelerated solvent extraction followed by solid-phase extraction. *Analytical Chemistry.* **74**(21): 5455-5462.

Golet, E.M., I. Xifra, H. Siegrist, A.C. Alder, and W. Giger (2003). Environmental exposure assessment of fluoroquinolone antibacterial agents from sewage to soil. *Environmental Science and Technology.* **37**(15): 3243-3249.

Goss, K.U. (2008). The pK(a) values of PFOA and other highly fluorinated carboxylic acids. *Environmental Science and Technology.* **42**(2): 456-458.

Goss, K.U., G. Bronner, T. Harner, H. Monika, and T.C. Schmidt (2006). The partition behavior of fluorotelomer alcohols and olefins. *Environmental Science and Technology.* **40**(11): 3572-3577.

Gouin, T. and T. Harner (2003). Modelling the environmental fate of the polybrominated diphenyl ethers. *Environment International.* **29**(6): 717-724.

Gu, C. and K.G. Karthikeyan (2005a). Interaction of tetracycline with aluminum and iron hydrous oxides. *Environmental Science and Technology.* **39**(8): 2660-2667.

Gu, C. and K.G. Karthikeyan (2005b). Sorption of the antimicrobial ciprofloxacin to aluminum and iron hydrous oxides. *Environmental Science and Technology.* **39**(23): 9166-9173.

Gu, C., K.G. Karthikeyan, S.D. Sibley and J.A. Pedersen (2007). Complexation of the antibiotic tetracycline with humic acid. *Chemosphere.* **66**(8): 1494-1501.

Guo, R., Q.F. Zhou, Y.Q. Cai, and G.B. Jiang (2008). Determination of perfluorooctanesulfonate and perfluorooctanoic acid in sewage sludge samples using liquid chromatography/ quadrupole time-of-flight mass spectrometry. *Talanta.* **75**(5): 1394-1399.

GWRC (2003). Endocrine Disrupting Compounds: Priority List of EDCs. London, United Kingdom, Global Water Research Coalition (GWRC).

Hale, R.C., M.J. La Guardia, E. Harvey, and T.M. Mainor (2002). Potential role of fire retardant-treated polyurethane foam as a source of brominated diphenyl ethers to the US environment, Pergamon-Elsevier Science Ltd.

Halling-Sorensen, B. (2001). Inhibition of aerobic growth and nitrification of bacteria in sewage sludge by antibacterial agents. *Archives of Environmental Contamination and Toxicology.* **40**: 451–460.

Halling-Sorensen, B., H.-C.H. Lutzhoft, H.R. Andersen, and F. Ingerslev (2000). Environmental risk assessment of antibiotics: comparison of mecillinam, trimethoprim and ciprofloxacin. *Journal of Antimicrobial Chemotherapy.* **46**(suppl_1): 53-58.

Hanselman, T.A., D.A. Graetz, and A.C. Wilkie (2003). Manure-borne estrogens as potential environmental contaminants: A review. *Environmental Science and Technology.* **37**(24): 5471-5478.

Harner, T. and M. Shoeib (2002). Measurements of octanol-air partition coefficients (K-OA) for polybrominated diphenyl ethers (PBDEs): Predicting partitioning in the environment. *Journal of Chemical and Engineering Data.* **47**(2): 228-232.

Harrison, E., S. Oakes, M. Hysell, and A. Hay (2006). Organic chemicals in sewage sludges. *Science of the Total Environment.* **367**(2-3): 481-497.

Hatzinger, P.B. and M. Alexander (1995). Effect of aging on chemicals in soil on their biodegradability and extractability. *Environmental Science and Technology.* **29**(2): 537-545.

He, J., K. Robrock, and L. Alvarez-Cohen (2006). Microbial reductive debromination of polybrominated diphenyl ethers (PBDEs). *Environmental Science and Technology.* **40**: 4429-4434.

Health Canada (2008). Health Risk Assessment of Bisphenol A From Food Packaging Applications. Ottawa, Ontario, Canada, Health Canada.

Heidler, J. and R.U. Halden (2007). Mass balance assessment of triclosan removal during conventional sewage treatment. *Chemosphere.* **66**(2): 362-369.

Hekster, F.M., R.W.P.M. Laane, and P. de Voogt (2003). Environmental and toxicity effects of perfluoroalkylated substances. *Reviews of Environmental Contamination and Toxicology.* **179**: 99-121.

Herren, D. and J.D. Berset (2000). Nitro musks, nitro musk amino metabolites and polycyclic musks in sewage sludges - Quantitative determination by HRGC-ion-trap-MS/MS and mass spectral characterization of the amino metabolites. *Chemosphere.* **40**(5): 565-574.

Higgins, C.P., J.A. Field, C.S. Criddle, and R.G. Luthy (2005). Quantitative determination of perfluorochemicals in sediments and domestic sludge. *Environmental Science and Technology.* **39**(11): 3946-3956.

Higgins, C.P. and R.G. Luthy (2006). Sorption of perfluorinated surfactants on sediments. *Environmental Science and Technology.* **40**(23): 7251-7256.

Higgins, C.P. and R.G. Luthy (2007). Modeling sorption of anionic surfactants onto sediment materials: An a priori approach for perfluoroalkyl surfactants and linear alkylbenzene sulfonates. *Environmental Science and Technology.* **41**(9): 3254-3261.

Higgins, C.P., P.B. McLeod, L.A. Macmanus-Spencer, and R.G. Luthy (2007). Bioaccumulation of perfluorochemicals in sediments by the aquatic oligochaete Lumbriculus variegatus. *Environmental Science and Technology.* **41**(13): 4600-4606.

Higgins, C.P., Z.J. Paesani, T.E.A. Chalew, and R.U. Halden (2009). Bioaccumulation of Triclocarban in Lumbriculus variegatus. *Environmental Toxicology and Chemistry.* **28**(12): 2580-2586.

Hildebrand, C., K.L. Londry, and A. Farenhorst (2006). Sorption and desorption of three endocrine disrupters in soils. *Journal of Environmental Science and Health Part B-Pesticides Food Contaminants and Agricultural Wastes.* **41**(6): 907-921.

Howdeshell, K.L., J. Furr, C.R. Lambright, V.S. Wilson, B.C. Ryan, and L.E. Gray, Jr. (2008). Gestational and lactational exposure to ethinyl estradiol, but not bisphenol A, decreases androgen-dependent reproductive organ weights and epididymal sperm abundance in the male Long Evans hooded rat. *Toxicological Sciences.* **102**(2): 371-382.

Hseu, Z.-Y. (2006). Response of microbial activities in two contrasting soils to 4-nonylphenol treated with biosolids. *Chemosphere.* **64**: 1769-1776.

Hydromantis. (2009a). TOXCHEM+: Modeling analysis tool for the fate of toxic in wastewater treatment. from http://www.hydromantis.com/TOXCHEM.html.

Hydromantis, U.o.W., Trent University, (2009b). Review of State of Knowledge of Biosolids Science and Research: Contaminant Inventory.

IEH (2005). Chemicals Purported to be Endocrine Disruptors: a Compilation of Published Lists (Web Report W20). Leicester, United Kingdom, Medical Research Council (MRC), Institute for Environment and Health (IEH).

ILSI. (2006). Summary of Evaluations Performed by the Joint FAO/WHO Expert Committee on Food Additives (JECFA 1956-2005). First through sixty-fifth meetings. Internet Edition Published by ISLI Press. International Life Sciences Institute. Washington, D.C. from http://jecfa.ilsi.org/.

Jacobs, L.W., G. O'Connor, M.A. Overcash, M.J. Zabik, and P. Rygiewicz (1987). Land application of sludge: food chain implications.*In* Effects of trace organics in sewate sludges on soil-plant systems and assessing their risk to humans. Chelsea, Michigan, Lewis Publishers, Inc.: 101-143.

Jacobsen, A., A. Lorenzen, R. Chapman, and E. Topp (2005). Persistence of testosterone and 17beta-estradiol in soils receiving swine manure or municipal biosolids. *Journal of Environmental Quality.* **34**(3): 861-71.

JECFA (1998). Tetracycline: Summary of Evaluations Performed by the Joint FAO/WHO Expert Committe on Food Additives (JECFA), Joint FAO/WHO Expert Committe on Food Additives (JECFA).

JECFA (2005). Polybrominated Diphenyl Ethers: Summary of Evaluations Performed by the Joint FAO/WHO Expert Committe on Food Additives (JECFA), Joint FAO/WHO Expert Committe on Food Additives (JECFA).

Johnson, A.C., C. White, T.J. Besien, and M.D. Jurgens (1998). The sorption potential of octylphenol, a xenobiotic oestrogen, to suspended and bed-sediments collected from industrial and rural reaches of three English rivers. *Science of the Total Environment.* **210**(1-6): 271-282.

Johnson, R.L., A.J. Anschutz, J.M. Smolen, M.F. Simcik, and R.L. Penn (2007). The adsorption of perfluorooctane sulfonate onto sand, clay, and iron oxide surfaces. *Journal of Chemical and Engineering Data.* **52**(4): 1165-1170.

Jones-Lepp, T.L. and R. Stevens (2007). Pharmaceuticals and personal care products in biosolids/sewage sludge: the interface between analytical chemistry and regulation. *Analytical and Bioanalytical Chemistry.* **387**(4): 1173-1183.

Jones, K. and J. Stevens (2002). Organic contaminants in sewage sludge applied to agricultural land. A critical evaluation of the proposed limit values for organics in the EU Working Document on Sludge and development of a tiered screening processes to identify priority pollutants in sewage sludge. UKWIR Report.

Jones, O.A., N. Voulvoulis, and J.N. Lester (2002). Aquatic environmental assessment of the top 25 English prescription pharmaceuticals. *Water Research* **36**(20): 5013-5022.

Kaleta, A., M. Ferdig, and W. Buchberger (2006). Semiquantitative determination of residues of amphetamine in sewage sludge samples. *Journal of Separation Science.* **29**(11): 1662-1666.

Katayama, A., R. Bhula, G.R. Burns, E. Carazo, A. Felsot, D. Hamilton, C. Harris, Y.H. Kim, G. Kleter, W. Koedel, J. Linders, J.G.M.W. Peijnenburg, A. Sabljic, R.G. Stephenson, D.K. Racke, B. Rubin, K. Tanaka, J. Unsworth, and R.D. Wauchope (2010). Bioavailability of Xenobiotics in the Soil Environment. *Reviews of Environmental Contamination and Toxicology.* **203**: 1-86.

Kawaguchi, M., K. Inoue, M. Yoshimura, R. Ito, N. Sakui, and H. Nakazawa (2004). Determination of 4-nonylphenol and 4-tert-octylphenol in water samples by stir bar sorptive extraction and thermal desorption gas chromatography mass spectrometry. *Analytica Chimica Acta.* **505**(2): 217-222.

Khan, M. and J. Scullion (1999). Microbial activity in grassland soil amended with sewage sludge containing varying rates and combinations of Cu, Ni and Zn. *Biology and Fertility of Soils.* **30**(3): 202-209.

Kidd, K.A., P.J. Blanchfield, K.H. Mills, V.P. Palace, R.E. Evans, J.M. Lazorchak, and R.W. Flick (2007). Collapse of a fish population after exposure to a synthetic estrogen. *Proceedings of the National Academy of Sciences of the United States of America.* **104**(21): 8897-8901.

Kim, S., P. Eichhorn, J.N. Jensen, A.S. Weber, and D.S. Aga (2005). Removal of antibiotics in wastewater: Effect of hydraulic and solid retention times on the fate of tetracycline in the activated sludge process. *Environmental Science and Technology.* **39**(15): 5816-5823.

Kinney, C., E. Furlong, D. Kolpin, M. Burkhardt, S. Zaugg, S. Werner, J. Bossio, and M. Benotti (2008). Bioaccumulation of pharmaceuticals and other anthropogenic waste indicators in earthworms from agricultural soil amended with biosolid or swine manure. *Environmental Science and Technology.* **42**(6): 1863-1870.

Kinney, C.A., E.T. Furlong, S.D. Zaugg, M.R. Burkhardt, S.L. Werner, J.D. Cahill, and G.R. Jorgensen (2006). Survey of organic wastewater contaminants in biosolids destined for land application. *Environmental Science and Technology.* **40**(23): 7207-7215.

Kong, W.D., Y.G. Zhu, B.J. Fu, P. Marschner, and J.Z. He (2006). The veterinary antibiotic oxytetracycline and Cu influence functional diversity of the soil microbial community. *Environmental Pollution.* **143**(1): 129-37.

Kong, W.D., Y.G. Zhu, Y.C. Liang, J. Zhang, F.A. Smith, and A. Yang (2007). Uptake of oxytetracycline and its phytotoxicity to alfalfa (Medicago sativa L.). *Environmental Pollution.* **147**(1): 187-193.

Kuch, H.M. and K. Ballschmiter (2001). Determination of endocrine-disrupting phenolic compounds and estrogens in surface and drinking water by HRGC-(NCI)-MS in the picogram per liter range. *Environmental Science and Technology.* **35**(15): 3201-3206.

Kuiper, R.V., R.F. Canton, P.E.G. Leonards, B.M. Jenssen, M. Dubbeldam, P.W. Wester, M. van den Berg, J.G. Vos, and A.D. Vethaak (2007). Long-term exposure of European flounder (Platichthys flesus) to the flame-retardants tetrabromobisphenol A (TBBPA) and hexabromocyclododecane (HBCD). *Ecotoxicology and Environmental Safety.* **67**(3): 349-360.

Kulshrestha, P., R.F. Giese, and D.S. Aga (2004). Investigating the molecular interactions of oxytetracycline in clay and organic matter: Insights on factors affecting its mobility in soil. *Environmental Science and Technology.* **38**(15): 4097-4105.

Kumar, K., S.C. Gupta, S.K. Baidoo, Y. Chander, and C.J. Rosen (2005). Antibiotic uptake by plants from soil fertilized with animal manure. *Journal of Environmental Quality.* **34**(6): 2082-2085.

Kupper, T., J.D. Berset, R. Etter-Holzer, R. Furrer, and J. Tarradellas (2004). Concentration and specific loads of polycyclic musks in sewage sludge originating from a monitoring network in Switzerland. *Chemosphere.* **54**(8): 1111-1120.

Kuramochi, H., K. Maeda, and K. Kawamoto (2007). Physicochemical properties of selected polybrominated diphenyl ethers and extension of the UNIFAC model to brominated aromatic compounds, Pergamon-Elsevier Science Ltd.

Laegdsmand, M., H. Andersen, O.H. Jacobsen, and B. Halling-Sorensen (2009). Transport and Fate of Estrogenic Hormones in Slurry-treated Soil Monoliths. *Journal of Environmental Quality.* **38**(3): 955-964.

Lai, K.M., K.L. Johnson, M.D. Scrimshaw, and J.N. Lester (2000). Binding of waterborne steroid estrogens to solid phases in river and estuarine systems. *Environmental Science and Technology.* **34**(18): 3890-3894.

Lapen, D., E. Topp, M. Edwards, L. Sabourin, W. Curnoe, N. Gottschall, P. Bolton, S. Rahman, B. Ball-Coelho, M. Payne, S. Kleywegt, and N. McLaughlin (2008a). Effect of liquid municipal biosolid application method on tile and ground water quality. *Journal of Environmental Quality.* **37**(3): 925-36.

Lapen, D.R., E. Topp, C.D. Metcalfe, H. Li, M. Edwards, N. Gottschall, P. Bolton, W. Curnoe, M. Payne, and A. Beck (2008b). Pharmaceutical and personal care products in tile drainage following land application of municipal biosolids. *Science of the Total Environment.* **399**(1-3): 50-65.

Law, R.J., M. Alaee, C.R. Allchin, J.P. Boon, M. Lebeuf, P. Lepom, and G.A. Stern (2003). Levels and trends of polybrominated diphenylethers and other brominated flame retardants in wildlife. *Environment International.* **29**(6): 757-770.

Lawlor, K., B.P. Knight, V.L. Barbosa-Jefferson, P.W. Lane, A.K. Lilley, G.I. Paton, S.P. McGrath, S.M. O'Flaherty, and P.R. Hirsch (2000). Comparison of methods to investigate microbial populations in soils under different agricultural management. *Fems Microbial Ecology.* **33**(2): 129-137.

Lawrence, J., B. Zhu, G. Swerhone, J. Roy, L. Wassenaar, E. Topp, and D. Korber (2009). Comparative microscale analysis of the effects of triclosan and triclocarban on the structure and function of river biofilm communities. *Science of the Total Environment.* **407**(10): 3307-16.

Layton, A.C., B.W. Gregory, J.R. Seward, T.W. Schultz, and G.S. Sayler (2000). Mineralization of Steroidal Hormones by Biosolids in Wastewater Treatment Systems in Tennessee U.S.A. *Environmental Science and Technology.* **34**(18): 3925-3931.

Lee, H.B. and T.E. Peart (2002). Organic contaminants in Canadian municipal sewage sludge. Part I. Toxic or endocrine-disrupting phenolic compounds. *Water Quality Research Journal of Canada.* **37**(4): 681-696.

Lee, L.S., N. Carmosini, S.A. Sassman, H.M. Dion, and M.S. Sepulveda (2007). Agricultural contributions of antimicrobials and hormones on soil and water quality. *Advances in Agronomy, Vol 93.* **93**: 1-68.

Lee, L.S., T.J. Strock, A.K. Sarmah, and P.S.C. Rao (2003). Sorption and dissipation of testosterone, estrogens, and their primary transformation products in soils and sediment. *Environmental Science and Technology.* **37**(18): 4098-4105.

Lei, Y.D., F. Wania, D. Mathers, and S.A. Mabury (2004). Determination of vapor pressures, octanol-air, and water-air partition coefficients for polyfluorinated sulfonamide, sulfonamidoethanols, and telomer alcohols. *Journal of Chemical and Engineering Data.* **49**(4): 1013-1022.

Lemos, M.F.L., C.A.M. van Gestel, and A. Soares (2009). Endocrine disruption in a terrestrial isopod under exposure to bisphenol A and vinclozolin. *Journal of Soils and Sediments.* **9**(5): 492-500.

Liebig, M., P. Egeler, J. Oehlmann, and T. Knacker (2005). Bioaccumulation of C-14-17 alpha-ethinylestradiol by the aquatic oligochaete Lumbriculus variegatus in spiked artificial sediment. *Chemosphere.* **59**(2): 271-280.

Lijzen, J. and M. Rikken (2004). 2004. EUSES 2.0 Background Report. Bilthoven, The Netherlands, Nation Institute of Public Health and the Environment. RIVM report no. 601900005.

Litz, N. (2002). Some investigations into the behavior of pentabromodiphenyl ether (PeBDE) in soils. *Journal of Plant Nutrition and Soil Science-Zeitschrift Fur Pflanzenernahrung Und Bodenkunde.* **165**(6): 692-696.

Litz, N.T., J. Muller, and W. Bohmer (2007). Occurrence of polycyclic musks in sewage sludge and their behaviour in soils and plants - Part 2: Investigation of polycyclic musks in soil and plants. *Journal of Soils and Sediments.* **7**(1): 36-44.

Liu, F., G.-G. Ying, L.-H. Yang, and Q.-X. Zhou (2009). Terrestrial ecotoxicological effects of the antimicrobial agent triclosan. *Ecotoxicology and Environmental Safety.* **72**(1): 86-92.

Liu, J., L.S. Lee, L.F. Nies, C.H. Nakatsu, and R.F. Turcot (2007). Biotransformation of 8:2 fluorotelomer alcohol in soil and by soil bacteria isolates. *Environmental Science and Technology.* **41**(23): 8024-30.

Liu, J.X. and L.S. Lee (2005). Solubility and sorption by soils of 8 : 2 fluorotelomer alcohol in water and cosolvent systems. *Environmental Science and Technology.* **39**(19): 7535-7540.

Liu, J.X. and L.S. Lee (2007). Effect of fluorotelomer alcohol chain length on aqueous solubility and sorption by soils. *Environmental Science and Technology.* **41**(15): 5357-5362.

Lobos, J., T. Leib, and T. Su (1992). Biodegradation of bisphenol A and other bisphenols by a gram-negative aerobic bacterium. *Applied and Environmental Microbiology.* **58**(6): 1823-31.

Loganathan, B.G., K.S. Sajwan, E. Sinclair, K. Senthil Kumar, and K. Kannan (2007). Perfluoroalkyl sulfonates and perfluorocarboxylates in two wastewater treatment facilities in Kentucky and Georgia. *Water Research.* **41**(20): 4611-4620.

Loke, M.L., J. Tjornelund, and B. Halling-Sorensen (2002). Determination of the distribution coefficient (log K-d) of oxytetracycline, tylosin A, olaquindox and metronidazole in manure. *Chemosphere.* **48**(3): 351-361.

Lozano, N., C.P. Rice, M. Ramirez, and A. Torrents (in press). Fate of triclosan in agricultural soils after biosolid applications. *Chemosphere.*

Luthy, R.G., G.R. Aiken, M.L. Brusseau, S.D. Cunningham, P.M. Gschwend, J.J. Pignatello, M. Reinhard, S.J. Traina, W.J. Weber, and J.C. Westall (1997). Sequestration of hydrophobic organic contaminants by geosorbents. *Environmental Science and Technology.* **31**(12): 3341-3347.

MacKay, A.A. and B. Canterbury (2005). Oxytetracycline sorption to organic matter by metal-bridging. *Journal of Environmental Quality.* **34**(6): 1964-1971.

Mackay, A.A. and D.E. Seremet (2008). Probe Compounds to Quantify Cation Exchange and Complexation Interactions of Ciprofloxacin with Soils. *Environmental Science and Technology.* **42**(22): 8270-8276.

Matscheko, N., M. Tysklind, C. de Wit, S. Bergek, R. Andersson, and U. Sellstrom (2002). Application of sewage sludge to arable land-soil concentrations of polybrominated diphenyl ethers and polychorinated dibenzo-p-dioxins, dibenzofurans, and biphenyls, and their accumulation in earthworms. *Environmental Toxicology and Chemistry.* **21**(12): 2515-2525.

McAvoy, D., W. Schecher, B. Rittmann, and K. Lee (1999). ASTREAT: A model for Calculating Chemical Loss Within an Activated Sludge System, User's Manual, Version 1.0. Cincinnati, OH, The Procter & Gamble Company.

McBain, A., R. Bartolo, C. Catrenich, D. Charbonneau, R. Ledder, B. Price, and P. Gilbert (2003). Exposure of sink drain microcosms to triclosan: population dynamics and antimicrobial susceptibility. *Applied and Environmental Microbiology.* **69**(9): 5433-42.

McDonald, T. (2005). Polybrominated diphenylether levels among United States residents: daily intake and risk of harm to the developing brain and reproductive organs. *Integrated Environmental Assessment and Management.* **1**(4): 343-354.

Mid-Continent Ecology Division. Duluth, Minnesota from <http://cfpub.epa.gov/ecotox/ecotox_home.cfm>.

Minnesota Department of Health (2005). 3M Chemolite Perfluorochemical Releases at the 3M-Cottage Grove Facility, City of Cottage Grove, Washington County, Minnesota. St. Paul, Minnesota, Prepared by the Minnesota Department of Health, under cooperative agreement with the United States Department of Health and Human Services, Agency for Toxic Substances and Disease Registry.

Minnesota Department of Health (2008). Health Risk Limits for Perfluorochemicals: Report to the Minnesota Legislature 2008 (Final Report). January 15. St. Paul, Minnesota, Minnesota Department of Health.

Minnesota Department of Health (2009a). Health Risk Limits for Groundwater 2008 Rule Revision: Perfluorooctane Sulfonate (PFOS). St. Paul, Minnesota, Minnesota Department of Health, Health Risk Assessment Unit, Environmental Health Division.

Minnesota Department of Health (2009b). Health Risk Limits for Groundwater 2008 Rule Revision: Perfluorooctanoic Acid (PFOA). St. Paul, Minnesota, Minnesota Department of Health, Health Risk Assessment Unit, Environmental Health Division.

Moran, M.A., V.L. Torsvik, T. Torsvik, and R.E. Hodson (1993). Direct extraction and purification of ribosomal-RNA for ecological-studies. *Applied and Environmental Microbiology.* **59**(3): 915-918.

Morris, S., C.R. Allchin, B.N. Zegers, J.J.H. Haftka, J.P. Boon, C. Belpaire, P.E.G. Leonards, S.P.J. Van Leeuwen, and J. De Boer (2004). Distributon and fate of HBCD and TBBPA brominated flame retardants in north sea estuaries and aquatic food webs. *Environmental Science and Technology.* **38**(21): 5497-5504.

Mueller, K.E., S.R. Mueller-Spitz, H.F. Henry, A.P. Vonderheide, R.S. Soman, B.K. Kinkle, and J.R. Shann (2006). Fate of pentabrominated diphenyl ethers in soil: Abiotic sorption, plant uptake, and the impact of interspecific plant interactions. *Environmental Science and Technology.* **40**(21): 6662-6667.

Mumma, R.O., D.C. Raupach, J.P. Waldman, S.S.C. Tong, M.L. Jacobs, J.G. Babish, J.H. Hotchkiss, P.C. Wszolek, W.H. Gutenman, C.A. Bache, and D.J. Lisk (1984). National survey of elements and other constituents in municipal sewage sludges. *Archives of Environmental Contamination and Toxicology.* **13**(1): 75-83.

Nakazawa, T., K. Kasahara, S. Ikezaki, Y. Yamaguchi, H. Edamoto, N. Nishimura, M. Yahata, K. Tamura, E. Kamata, M. Ema, and R. Hasegawa (2009). Renal tubular cyst formation in newborn rats treated with p-cumylphenol. *Journal of Toxicologic Pathology.* **22**(2): 125-131.

Nalli, S., D.G. Cooper, and J.A. Nicell (2006). Interaction of metabolites with R. rhodochrous during the biodegradation of di-ester plasticizers. *Chemosphere.* **65**(9): 1510-1517.

NAS (1983). Risk Assessment in the Federal Government: Managing the Process. Washington, D.C. National Academy Press.

Navarro, A., S. Endo, T. Gocht, J.A.C. Barth, S. Lacorte, D. Barcelo, and P. Grathwohl (2009). Sorption of alkylphenols on Ebro River sediments: Comparing isotherms with field observations in river water and sediments. *Environmental Pollution.* **157**(2): 698-703.

Newsted, J.L., P.D. Jones, K. Coady, and J.P. Giesy (2005). Avian toxicity reference values for perfluorooctane sulfonate. *Environmental Science and Technology.* **39**(23): 9357-9362.

Nicholls, C.R., C.R. Allchin, and R.J. Law (2001). Levels of short and medium chain length polychlorinated n-alkanes in environmental samples from selected industrial areas in England and Wales. *Environmental Pollution.* **114**(3): 415-430.

NIH. (2009). Hazardous Substances Data Bank (HSDB®). Accessed via the National Library of Medicine's Toxicology Data Nework (TOXNET®). National Institutes of Health, Health & Human Medicine. Bethesda, Maryland. from http://toxnet.nlm.nih.gov.

Norén, K. and D. Meironyté (2000). Certain organochlorine and organobromine contaminants in Swedish human milk in perspective of past 20-30 years. *Chemosphere.* **40**(9-11): 1111-1123.

Nowara, A., J. Burhenne, and M. Spiteller (1997). Binding of fluoroquinolone carboxylic acid derivatives to clay minerals. *Jorunal of Agricultural and Food Chemistry.* **45**(4): 1459-1463.

NRC (2002). Biosolids Applied to Land: Advancing Standards and Practices. Washington, D.C. National Academy Press.

NTP-CERHR (2007). NTP-CERHR Expert Panel Report on the Reproductive and Developmental Toxicity of Bisphenol A. Research Triangle Park, NC, National Toxicology Program (NTP), United States Department of Health and Human Services. NTP-CERHR-BPA-07.

NTP (2001). National Toxicology Program's Report of the Endocrine Disruptors Low-Dose Peer Review. Research Triangle Park, North Carolina, National Toxicology Program (NTP).

O'Connor, G.A. (1996). Organic compounds in sludge-amended soils and their potential for uptake by crop plants, Elsevier Science Bv.

Orvos, D.R., D.J. Versteeg, J. Inauen, M. Capdevielle, A. Rothenstein, and V. Cunningham (2002). Aquatic toxicity of triclosan. *Environmental Toxicology and Chemistry.* **21**(7): 1338-1349.

Palm, A., I.T. Cousins, D. Mackay, M. Tysklind, C. Metcalfe, and M. Alaee (2002). Assessing the environmental fate of chemicals of emerging concern: a case study of the polybrominated diphenyl ethers. *Environmental Pollution.* **117**(2): 195-213.

Pan, B., P. Ning, and B.S. Xing (2009). Part V-sorption of pharmaceuticals and personal care products. *Environmental Science and Pollution Research.* **16**(1): 106-116.

Parrott, J.L. and B.R. Blunt (2005). Life-cycle exposure of fathead minnows (Pimephales promelas) to an ethinylestradiol concentration below 1 ng/L reduces egg fertilization success and demasculinizes males. *Environmental Toxicology.* **20**(2): 131-141.

Petrovic, M. and D. Barcelo (2000). Determination of anionic and nonionic surfactants, their degradation products, and endocrine-disrupting compounds in sewage sludge by liquid chromatography/mass spectrometry. *Analytical Chemistry.* **72**(19): 4560-4567.

Pils, J.R.V. and D.A. Laird (2007). Sorption of tetracycline and chlortetracycline on K- and Ca-saturated soil clays, humic substances, and clay-humic complexes. *Environmental Science and Technology.* **41**(6): 1928-1933.

Poiger, T., J.A. Field, T.M. Field, H. Siegrist, and W. Giger (1998). Behavior of fluorescent whitening agents during sewage treatment. *Water Research.* **32**(6): 1939-1947.

Posthuma, L., G. Suter, and T. Traas, Eds. (2002). Species Sensitivity Distributions in Ecotoxicology. Boca Raton, FL, CRC Lewis Publishers.

Pruden, A., R. Pei, H. Storteboom, and K.H. Carlson (2006). Antibiotic Resistance Genes as Emerging Contaminants: Studies in Northern Colorado. *Environmental Science and Technology.* **40**(23): 7445-7450.

Qiang, Z.M. and C. Adams (2004). Potentiometric determination of acid dissociation constants (pK(a)) for human and veterinary antibiotics. *Water Research.* **38**(12): 2874-2890.

Rabolle, M. and N.H. Spliid (2000). Sorption and mobility of metronidazole, olaquindox, oxytetracycline and tylosin in soil. *Chemosphere.* **40**(7): 715-722.

Radjenovic, J., M. Petrovic, and D. Barcelo (2009). Fate and distribution of pharmaceuticals in wastewater and sewage sludge of the conventional activated sludge (CAS) and advanced membrane bioreactor (MBR) treatment. *Water Research.* **43**(3): 831-841.

Radl, V., K. Pritsch, J.C. Munch, and M. Schloter (2005). Structural and functional diversity of microbial communities from a lake sediment contaminated with trenbolone, an endocrine-disrupting chemical. *Environmental Pollution.* **137**(2): 345-53.

Ramskov, T., H. Selck, D. Salvito, and V.E. Forbes (2009). Individual- and population-level effects of the synthetic musk, HHCB, on the deposit-feeding polychaete, *Capitella sp. Environmental Toxicology and Chemistry.* **28**(12): 2695-2705.

Reiss, R., G. Lewis, and J. Griffin (2009). An ecological risk assessment for triclosan in the terrestrial environment. *Environmental Toxicology and Chemistry.* **28**(7): 1546-1556.

Renner, R. (2009). EPA finds record PFOS, PFOA levels in Alabama grazing fields. *Environmental Science and Technology.* **43**(5): 1245-1246.

Robrock, K., P. Korytar, and L. Alvarez-Cohen (2008). Pathways for the anaerobic microbial debromination of polybrominated diphenyl ethers. *Environmental Science and Technology.* **42**(8): 2845–2852.

RTI (2005). Methodology for Predicting Cattle Biotransfer Factors. Prepared for the United States EPA Office of Solid Waste. EPA Contract No. 68-W-03-042.

RTI (2008). Technical Background Document: Update Biosolids Exposure and Hazard Assessment. Washington, D.C. Prepared for United States EPA Office of Water.

Sabourin, L., A. Beck, P.W. Duenk, S. Kleywegt, D.R. Lapen, H. Li, C. Metcalfe, M. Payne, and E. Topp (2009). Runoff of pharmaceuticals and personal care products following application of dewatered municipal biosolids to an agricultural field. *Science of the Total Environment.* In press.

Saez, M., P. de Voogt, and J.R. Parsons (2008). Persistence of perfluoroalkylated substances in closed bottle tests with municipal sewage sludge. *Environmental Science and Pollution Research.* **15**(6): 472-477.

Samsøe-Petersen, L., M. Winther-Nielsen, and T. Madsen (2003). Fate and Effects of Triclosan [Environmental Project No. 861], Danish Environmental Protection Agency.

Sassman, S.A. and L.S. Lee (2005). Sorption of three tetracyclines by several soils: Assessing the role of pH and cation exchange. *Environmental Science and Technology.* **39**(19): 7452-7459.

Saviozzi, A., A. Biasci, R. Riffaldi, and R. Levi-Minzi (1999). Long-term effects of farmyard manure and sewage sludge on some soil biochemical characteristics. *Biology and Fertility of Soils.* **30**(1-2): 100-106.

Schnaak, W., T. Kuchler, M. Kujawa, K.P. Henschel, D. Sussenbach, and R. Donau (1997). Organic contaminants in sewage sludge and their ecotoxicological significance in the agricultural utilization of sewage sludge. *Chemosphere.* **35**(1-2): 5-11.

Schowanek, D., R. Carr, H. David, P. Douben, J. Hall, H. Kirchmann, L. Patria, P. Sequi, S. Smith, and S. Webb (2004). A risk-based methodology for deriving quality standards for organic contaminants in sewage sludge for use in agriculture - Conceptual Framework. *Regulatory Toxicology and Pharmacology.* **40**(3): 227-251.

Schowanek, D., H. David, R. Francaviglia, J. Hall, H. Kirchmann, P.H. Krogh, N. Schraepen, S. Smith, and T. Wildemann (2007). Probabilistic risk assessment for linear alkylbenzene sulfonate (LAS) in sewage sludge used on agricultural soil. *Regulatory Toxicology and Pharmacology.* **49**(3): 245-259.

Schreurs, R.H.M.M., J. Legler, E. Artola-Garicano, T.L. Sinnige, P.H. Lanser, W. Seinen, and B. van der Burg (2004). In vitro and in vivo antiestrogenic effects of polycyclic musks in zebrafish. *Environmental Science and Technology.* **38**(4): 997 -1002.

Schroder, H.F. (2003). Determination of fluorinated surfactants and their metabolites in sewage sludge samples by liquid chromatography with mass spectrometry and tandem mass spectrometry after pressurised liquid extraction and separation on fluorine-modified reversed-phase sorbents. *Journal of Chromatography A.* **1020**(1): 131-151.

Schwab, B.W., E.P. Hayes, J.M. Fiori, F.J. Mastrocco, N.M. Roden, D. Cragin, R.D. Meyerhoff, V.J. D'Aco, and P.D. Anderson (2005). Human pharmaceuticals in US surface waters: A human health risk assessment. *Regulatory Toxicology and Pharmacology.* **42**(3): 296-312.

Schwarzenbach, R., P. Gschwend, and D. Imboden (2003). Environmental Organic Chemistry. Hoboken, NJ, Wiley Interscience.

Sellstrom, U., C. De Wit, N. Lundgren, and M. Tysklind (2005). Effect of sewage-sludge application on concentrations of higher-brominated diphenyl ethers in soils and earthworms. *Environmental Science and Technology.* **39**(23): 9064-9070.

Sellstrom, U. and B. Jansson (1995). Analysis of tetrabromobisphenol a in a product and environmental-samples. *Chemosphere.* **31**(4): 3085-3092.

SETAC. (2009). Society of Environmental Toxicology and Chemistry (SETAC) journals including *Environmental Toxicology and Chemistry* and *Integrated Environmental Assessment and Management* from http://www.setac.org.

Shareef, A., M.J. Angove, J.D. Wells, and B.B. Johnson (2006). Sorption of bisphenol A, 17 alpha-ethynylestradiol and estrone to mineral surfaces. *Journal of Colloid and Interface Science.* **297**(1): 62-69.

Sharvelle, S., R. Lattyak, and a.M.K. Banks (2007). Evaluation of Biodegradability and Biodegradation Kinetics for Anionic, Nonionic, and Amphoteric Surfactants. *Water, Air, and Soil Pollution.* **183**: 177-186.

Shoeib, M., T. Harner, and P. Vlahos (2006). Perfluorinated chemicals in the Arctic atmosphere. *Environmental Science and Technology.* **40**(24): 7577-7583.

Sinclair, E. and K. Kannan (2006). Mass loading and fate of perfluoroalkyl surfactants in wastewater treatment plants. *Environmental Science and Technology.* **40**(5): 1408-1414.

Sithole, B.B. and R.D. Guy (1987a). Models for tetracycline in aquatic environments .1. Interaction with bentonite clay systems. *Water Air and Soil Pollution.* **32**(3-4): 303-314.

Sithole, B.B. and R.D. Guy (1987b). Models for tetracycline in aquatic environments .2. Interaction with humic substances. *Water Air and Soil Pollution.* **32**(3-4): 315-321.

Smith, S. (2008). The Implications for Human Health and the Environment of Recycling Biosolids on Agricultural Land, Imperial College London.

Smith, S.R. (2009). Organic contaminants in sewage sludge (biosolids) and their significance for agricultural recycling. *Philosophical Transactions of the Royal Society a-Mathematical Physical and Engineering Sciences.* **367**(1904): 4005-4041.

Smyth, S.A., L. Lishman, E.A. McBean, S. Kleywegt, J.J. Yang, M.L. Svoboda, H. Lee, and P. Seto (2007). Fate of Polycyclic and Nitro Musks during Aerobic and Anaerobic Sludge Digestion. International Water Association Specialist Conference on Moving Forward Biosolids Sustainability. Moncton NB.

Snyder, E. (2009). Fate, Transport, and Risk Assessment of Biosolids-Borne Triclocarban, University of Florida. Doctor of Philosophy: 237.

Snyder, S.A., R.A. Trenholm, E.M. Snyder, G.M. Bruce, R.C. Pleus, and J.D.C. Hemming (2008a). Toxicological Relevance of EDCs and Pharmaceuticals in Drinking Water. Denver, Colorado, American Water Works Association Research Foundation (AwwaRF).

Snyder, S.A., B.J. Vanderford, J. Drewes, E. Dickenson, E.M. Snyder, G.M. Bruce, and R.C. Pleus (2008b). State of Knowledge of Endocrine Disruptors and Pharmaceuticals in Drinking Water. Denver, Colorado. American Water Works Association Research Foundation (AwwaRF).

Spongberg, A.L. and J.D. Witter (2008). Pharmaceutical compounds in the wastewater process stream in Northwest Ohio. *Science of the Total Environment.* **397**(1-3): 148-157.

Stasinakis, A.S., A.V. Petalas, D. Mamais, and N.S. Thomaidis (2008). Application of the OECD 301F respirometric test for the biodegradability assessment of various potential endocrine disrupting chemicals. *Bioresource Technology.* **99**(9): 3458-3467.

Stasinakis, A.S., A.V. Petalas, D. Mamais, N.S. Thomaidis, G. Gatidou, and T.D. Lekkas (2007). Investigation of triclosan fate and toxicity in continuous-flow activated sludge systems. *Chemosphere.* **68**(2): 375-381.

Stock, N.L., V.I. Furdui, D.C.G. Muir, and S.A. Mabury (2007). Perfluoroalkyl contaminants in the canadian arctic: Evidence of atmospheric transport and local contamination. *Environmental Science and Technology.* **41**(10): 3529-3536.

Stomperudhaugen, E.S., N.H.H. Overas, K. Langford, W. de Coen, R. Smolders, and K. Hylland (2009). Cellular Energy Allocation in Hediste diversicolor Exposed to Sediment Contaminants. *Journal of Toxicology and Environmental Health-Part a-Current Issues.* **72**(3-4): 244-253.

Struijs, J. (1996). SimpleTreat 3.0: a model to predict the distribution and elimination of chemicals by sewage treatment plants. . Bilthoven, The Netherlands, Nation Institute of Public Health and the Environment. RIVM report no. 719101025. .

Stumpe, B. and B. Marschner (2009). Factors controlling the biodegradation of 17 beta-estradiol, estrone and 17 alpha-ethinylestradiol in different natural soils. *Chemosphere.* **74**(4): 556-562.

Sun, Y.Y., H.Y. Guo, H.X. Yu, X.R. Wang, J.C. Wu, and Y.Q. Xue (2008a). Bioaccumulation and physiological Effects of tetrabromobisphenol A in coontail Ceratophyllum demersum L. *Chemosphere.* **70**(10): 1787-1795.

Sun, Z.H., L. Mao, Q.M. Xian, Y.J. Yu, H. Li, and H.X. Yu (2008b). Effects of dissolved organic matter from sewage sludge on sorption of tetrabromobisphenol A by soils. *Journal of Environmental Sciences-China.* **20**(9): 1075-1081.

Sun, Z.H., Y.J. Yu, L. Mao, Z. Feng, and H.X. Yu (2008c). Sorption behavior of tetrabromobisphenol A in two soils with different characteristics. *Journal of Hazardous Materials.* **160**(2-3): 456-461.

Suter, G. (1993). A framework for assessing ecological risks of petroleum-derived materials in soil. Oak Ridge, TN. Environmental Sciences Division Publication No. 4666. Oak Ridge National Laboratory.

Sverdrup, L.E., T. Hartnik, E. Mariussen, and J. Jensen (2006). Toxicity of three halogenated flame retardants to nitrifying bacteria, red clover (Trifolium pratense), and a soil invertebrate (Enchytraeus crypticus). *Chemosphere.* **64**(1): 96-103.

Tan, B.L., D.W. Hawker, J.F. Muller, F.D. Leusch, L.A. Tremblay, and H.F. Chapman (2007). Comprehensive study of endocrine disrupting compounds using grab and passive sampling at selected wastewater treatment plants in South East Queensland, Australia. *Environment International.* **33**(5): 654-69.

Tan, B.L.L., D.W. Hawker, J.F. Muller, L.A. Tremblay, and H.F. Chapman (2008). Stir bar sorptive extraction and trace analysis of selected endocrine disruptors in water, biosolids and sludge samples by thermal desorption with gas chromatography-mass spectrometry. *Water Research.* **42**(1-2): 404-412.

Tate, K.R. and D.S. Jenkinson (1982). Adenosine-triphosphate measurement in soil - an improved method. *Soil Biology and Biochemistry.* **14**(4): 331-335.

Ternes, T.A., N. Herrmann, M. Bonerz, T. Knacker, H. Siegrist, and A. Joss (2004). A rapid method to measure the solid-water distribution coefficient (K-d) for pharmaceuticals and musk fragrances in sewage sludge. *Water Research.* **38**(19): 4075-4084.

Teske, S.S. and R.G. Arnold (2008). Removal of natural and xeno-estrogens during conventionalwastewater treatment. *Reviews in Environmental Science and Biotechnology.* **7**: 107–124.

Thiele-Bruhn, S. (2003). Pharmaceutical antibiotic compounds in soils - a review. *Journal of Plant Nutrition and Soil Science-Zeitschrift Fur Pflanzenernahrung Und Bodenkunde.* **166**(2): 145-167.

Thiele-Bruhn, S. and I.C. Beck (2005). Effects of sulfonamide and tetracycline antibiotics on soil microbial activity and microbial biomass. *Chemosphere.* **59**(4): 457-65.

Thuens, S., A. Dreyer, R. Sturm, C. Temme, and R. Ebinghaus (2008). Determination of the octanol-air partition coefficients (K-OA) of fluorotelomer alcohols. *Journal of Chemical and Engineering Data.* **53**(1): 223-227.

Tittlemier, S.A., T. Halldorson, G.A. Stern, and G.T. Tomy (2002). Vapor pressures, aqueous solubilities, and Henry's law constants of some brominated flame retardants. *Environmental Toxicology and Chemistry.* **21**(9): 1804-1810.

Tolls, J. (2001). Sorption of veterinary pharmaceuticals in soils: A review. *Environmental Science and Technology.* **35**(17): 3397-3406.

Topp, E., J.G. Hendel, D.R. Lapen, and R. Chapman (2008a). Fate of the nonsteroidal anti-inflammatory drug naproxen in agricultural soil receiving liquid municipal biosolids. *Environmental Toxicology and Chemistry.* **27**(10): 2005-2010.

Topp, E., S.C. Monteiro, A. Beck, B.B. Coelho, A.B.A. Boxall, P.W. Duenk, S. Kleywegt, D.R. Lapen, M. Payne, L. Sabourin, H.X. Li, and C.D. Metcalfe (2008b). Runoff of pharmaceuticals and personal care products following application of biosolids to an agricultural field. *Science of the Total Environment.* **396**(1): 52-59.

Topp, E. and A. Starratt (2000). Rapid mineralization of the endocrine-disrupting chemical 4-nonylphenol in soil. *Environmental Toxicology and Chemistry.* **19**(2): 313-318.

Torslov, J., L. Samsoe-Peterson, J.O. Rasmussen, and P. Kristensen (1997). Use of waste products in agriculture: contamination level, environmental risk assessment and recommendations for quality criteria, Ministry of Environment and Energy, Denmark. Danish Environmental Protection Agency. Environmental Project No. 366.

Traina, S.J., N.J. Fendinger, D.C. McAvoy, K.M. Kerr, and S. Gupta (2002). Fate of polydimethylsilicone in biosolids-amended field plots. *Journal of Environmental Quality.* **31**(1): 247-255.

Trent University (2009). Biosolids Amended Soil Level IV (BASL-4), v 1.10. Peterborough, Ontario.

Tyl, R.W., C.B. Myers, M.C. Marr, D.R. Brine, P.A. Fail, J.C. Seely, and J.P. Van Miller (1999). Two-generation reproduction study with para-tert-octylphenol in rats. *Regulatory Toxicology and Pharmacology.* **30**(2): 81-95.

UKWIR. (2009). Toxicity Datasheets: Database of Chemical Toxicity, Environmental Fate, and Water Treatment. United Kingdom Water Industry Research Limited (UKWIR) and WRc-NSF Limited.

U.S. EPA-IRIS. (2009). Integrated Risk Information System (IRIS) [Online database]. United States Environmental Protection Agency, National Center for Environmental Assessment. from http://cfpub.epa.gov/ncea/iris/index.cfm.

U.S. EPA-OSWER (2009). Memorandum: The Toxicity of Perfluorooctanoic Acid (PFOA) and Perfluorooctane Sulfonate (PFOS). Washington, D.C., United States Environmental Protection Agency, Office of Solid Waste and Emergency Response (U.S. EPA-OSWER).

U.S. EPA-OW (2009). 2009 Edition of the Drinking Water Standards and Health Advisories. EPA 822-R-09-011. (Updated October 2009). Washington, D.C., United States Environmental Protection Agency, Office of Water. EPA 822-R-09-011.

U.S. EPA (1992). Technical Support Document for Land Application of Sewage Sludge. Washington, D.C. 822R93001a.

U.S. EPA (1993a). The standards for the use of disposal of sewage sludge. Final rules. EPA 822/Z-93-001. 40 CFR Parts 257, 403, and 503. *Federal Register*. **58**: 9248-9404.

U.S. EPA (1993b). Wildlife Exposure Factors Handbook. Volumes I and II. Washington, D.C. EPA/600/R-93/187. Office of Health and Environmental Assessment and Office of Research and Development.

U.S. EPA (1995). A Guide to the Biosolids Risk Assessments for the EPA Part 503 Rule. 832/B-93-005.

U.S. EPA (1996). Technical support document for the round two sewage sludge pollutants. EPA-822-R-96-003.

U.S. EPA (1998a). Guidelines for Ecological Risk Assessment (Final). Washington, D.C. EPA/630/R-95/002F. Risk Assessment Forum.

U.S. EPA (1998b). Methodology for Assessing Health Risks Associated with Multiple Pathways of Exposure to Combustor Emissions. Cincinnati, OH. EPA/600/R-98/137. National Center for Environmental Assessment.

U.S. EPA (2000). Methods for Measuring the Toxicity and Bioaccumulation of Sediment-associated Contaminants with Freshwater Invertebrates. EPA 600/R-99/064.

U.S. EPA (2003a). News Release (10/17/2003): EPA Makes Final Decison on Dioxin In Sewage Sludge used in Land Applications. Washington, D.C.

U.S. EPA (2003b). Technical Background Document for the Sewage Sludge Exposure and Hazard Screening Assessment. Washington, D.C. Document No. 822-B-03-001.

U.S. EPA. (2008a). EPI Suite. from <http://www.epa.gov/oppt/exposure/docs/episuitedl.htm>.

U.S. EPA (2008b). Reregistration Eligibility Decision and Risk Assessment for the Pesticidal Uses of Triclosan (current as of October 2008) [Fact Sheet]. Washington, D.C., United States Environmental Protection Agency, Office of Prevention, Pesticides and Toxic Substances (U.S. EPA-OPPTS).

U.S. EPA (2008c). Reregistration Eligibility Decision for Triclosan (List B, Case No. 2340). EPA 739-RO-8009. September. Washington, D.C., United States Environmental Protection Agency, Office of Prevention, Pesticides and Toxic Substances (U.S. EPA-OPPTS).

U.S. EPA. (2009a). from http://www.epa.gov/ncea/iris/subst/0356.htm.

U.S. EPA (2009b). Drinking Water Standards and Health Advisories (Updated October 2009). Washington, D.C. United States Environmental Protection Agency, Office of Water. EPA 822-R-09-011. EPA-822-R-96-003.

U.S. EPA. (2009c). ECOTOX database. United States Environmental Protection Agency, Office of Research and Development, National Health and Environmental Effects Research Laboratory (NHEERL),

U.S. EPA. (2009d). Integrated Risk Information System (IRIS). United States Environmental Protection Agency, National Center for Environmental Assessment. Washington, D.C. from <http://cfpub.epa.gov/ncea/iris/index.cfm>.

U.S. EPA (2009e). Memorandum: The Toxicity of Perfluorooctanoic Acid (PFOA) and Perfluorooctane Sulfonate (PFOS). Washington, D.C., United States Environmental Protection Agency, Office of Solid Waste and Emergency Response (U.S. EPA-OSWER).

U.S. EPA (2009f). Occurrence of Contaminants of Emerging Concern in Wastewater From Nine Publicly Owned Treatment Works. Washington, D.C. Office of Water. EPA-821-R-09-009.

U.S. EPA (2009g). Perfluorochemical (PFC) Contamination in Dalton, GA: Statement and Background. United States Environmental Protection Agengy Regioin 4.

U.S. EPA. (2009h). Priority Pollutants. from http://www.epa.gov/waterscience/methods/pollutants.htm.

U.S. EPA (2009i). Targeted National Sewage Sludge Survey. Washington, D.C., United States Environmental Protection Agency, Office of Water.

U.S. EPA. (2010). Glossary. from http://www.epa.gov/risk/glossary.htm.

USGS. (2009). Health-based screening levels (HBSLs). from http://infotrek.er.usgs.gov/traverse/f?p=169:4:0::NO::P4_WHICH_SECTION:ALL and http://infotrek.er.usgs.gov/traverse/f?p=169:30:0::NO:RP:P30_REPORT_TYPE:NONE.

Uslu, M.O., A. Yediler, I.A. Balcioglu, and S. Schulte-Hostede (2008). Analysis and sorption behavior of fluoroquinolones in solid matrices. *Water Air and Soil Pollution.* **190**(1-4): 55-63.

Vader, J.S., C.G. van Ginkel, F.M.G.M. Sperling, J. de Jong, W. de Boer, J.S. de Graaf, M. van der Most, and P.G.W. Stokman (2000). Degradation of ethinyl estradiol by nitrifying activated sludge. *Chemosphere.* **41**(8): 1239-1243.

Van de Meent, D., J. De Bruyn, F. Leeuw, A. Nijs, D. Jager, and T. Vermeire (1995). Exposure modeling. *In* Risk Assessment of Chemicals, an Introduction. C.J. van Leeuwen and C.J. Hermens. Dordrecht, Kluwer Academic Publishers.

Van de Plassche, E., and F. Balk (1997). Environmental risk assessment of the polycyclic musks AHTN and HHCB according to the EU-TGD. Bilthoven, The Netherlands, Nation Institute of Public Health and the Environment. RIVM Report 601503008.

Van Miller, J.P., and C.A. Staples (2005). Review of the potential environmental and human health-related hazards and risks from long-term exposure to p-tert-octylphenol. *Human and Ecological Risk Assessment.* **11**(2): 319.

van Wezel, A.P. and T. Jager (2002). Comparison of two screening level risk assessment approaches for six disinfectants and pharmaceuticals. *Chemosphere.* **47**(10): 1113-1128.

Vanden Heuvel, J. (2009). PPAR Resource Page. from http://ppar.cas.psu.edu/

Vasconcelos, T., D. Henriques, A. König, A. Martins, and K. Kümmerer (2009). Photo-degradation of the antimicrobial ciprofloxacin at high pH: identification and biodegradability assessment of the primary by-products. *Chemosphere.* **76**(4): 487-93.

Vasudevan, D., G.L. Bruland, B.S. Torrance, V.G. Upchurch, and A.A. MacKay (2009). pH-dependent ciprofloxacin sorption to soils: Interaction mechanisms and soil factors influencing sorption. *Geoderma.* **151**(3-4): 68-76.

Villeneuve, D.L., P. Larkin, I. Knoebl, A.L. Miracle, M.D. Kahl, K.M. Jensen, E.A. Makynen, E.J. Durhan, B.J. Carter, N.D. Denslow, and G.T. Ankley (2007). A graphical systems model to facilitate hypothesis-driven ecotoxicogenomics research on the teleost brain-pituitary-gonadal axis. *Environmental Science and Technology.* **41**(1): 321-330.

Vonderheide, A., S. Mueller-Spitz, J. Meija, Welsh, Gwendolyn, K. Mueller, B. Kinkle, J. Shann and J., and Caruso (2006). Rapid breakdown of brominated flame retardants by soil microorganisms. *Journal of Analytical Atomic Spectrometry.* **21**: 7.

Vonderheide, A.P., K.E. Mueller, J. Meija, and G.L. Welsh (2008). Polybrominated diphenyl ethers: Causes for concern and knowledge gaps regarding environmental distribution, fate and toxicity. *Science of the Total Environment.* **400**(1-3): 425-436.

Waller, N.J. and R.S. Kookana (2009). Effect of triclosan on microbial activity in Australian soils. *Environmental Toxicology and Chemistry.* **28**(1): 65-70.

Walsh, L.P., C.N. Kuratko, and D.M. Stocco (2000). Econazole and miconazole inhibit steroidogenesis and disrupt steroidogenic acute regulatory (StAR) protein expression post-transcriptionally. *Journal of Steroid Biochemistry and Molecular Biology.* **75**(4-5): 229-236.

Wang, N., B. Szostek, R.C. Buck, P.W. Folsom, L.M. Sulecki, V. Capka, W.R. Berti, and J.T. Gannon (2005a). Fluorotelomer alcohol biodegradation-direct evidence that perfluorinated carbon chains breakdown. *Environmental Science and Technology.* **39**(19): 7516-28.

Wang, N., B. Szostek, R.C. Buck, P.W. Folsom, L.M. Sulecki, and J.T. Gannon (2009). 8-2 Fluorotelomer alcohol aerobic soil biodegradation: Pathways, metabolites, and metabolite yields. *Chemosphere.* **75**(8): 1089-1096.

Wang, N., B. Szostek, P.W. Folsom, L.M. Sulecki, V. Capka, R.C. Buck, W.R. Berti, and J.T. Gannon (2005b). Aerobic biotransformation of C-14-labeled 8-2 telomer B alcohol by activated sludge from a domestic sewage treatment plant. *Environmental Science and Technology.* **39**(2): 531-538.

Wang, Y.J., D.A. Jia, R.J. Sun, H.W. Zhu, and D.M. Zhou (2008). Adsorption and cosorption of tetracycline and copper(II) on montmorillonite as affected by solution pH. *Environmental Science and Technology.* **42**(9): 3254-3259.

Wania, F. and C.B. Dugani (2003). Assessing the long-range transport potential of polybrominated diphenyl ethers: A comparison of four multimedia models. *Environmental Toxicology and Chemistry.* **22**(6): 1252-1261.

Washington, J.W., J.J. Ellington, T.M. Jenkins, J.J. Evans, H. Yoo, and S.C. Hafner (2009). Degradability of an Acrylate-Linked, Fluorotelomer Polymer in Soil. *Environmental Science and Technology.* **43**(17): 6617-6623.

Watanabe, I. and S. Sakai (2003). Environmental release and behavior of brominated flame retardants. *Environment International.* **29**(6): 665-682.

Watanabe, N., T. Nakamura, E. Watanabe, E. Sato, and Y. Ose (1984a). Distribution of organosiloxanes (silicones) in water, sediments and fish from the Nagara River watershed, Japan. *Science of the Total Environment.* **35**(1): 91-97.

Watanabe, N., Y. Yasuda, K. Kato, T. Nakamura, R. Funasaka, K. Shimokawa, E. Sato, and Y. Ose (1984b). Determination of trace amounts of siloxanes in water, sediments and fish-tissues by inductively coupled plasma emission-spectrometry. *Science of the Total Environment.* **34**(1-2): 169-176.

White, D.C. (1995). Chemical ecology: Possible linkage between macro- and microbial ecology. *Oikos.* **74**(2): 177-184.

WHO-IPCS (2002). Global Assessment of the State-of-the-Science of Endocrine Disruptors. Geneva, Switzerland, World Health Organization (WHO), International Programme on Chemical Safety (IPCS). WHO/PCS/EDC/02.2.

WHO. (2000). Toxicological Evaluation of Certain Veterinary Drug Residues in Food. WHO Food Additive Series: 43. Prepared by the fifty-second meeting of the Joint FAO/WHO Expert Committee on Food Additives (JECFA). World Health Organization, International Programme on Chemical Safety (IPCS). Geneva, Switzerland. from http://www.inchem.org/documents/jecfa/jecmono/v43jec05.htm.

WHO. (2009). Concise International Chemical Assessment Documents (CICADs). The International Programme on Chemical Safety (IPCS). Geneva, Switzerland. from http://www.who.int/ipcs/publications/cicad/en/index.html.

Wong, C.S., P.D. Capel, and L.H. Nowell (2001). National-scale, field-based evaluation of the biota-sediment accumulation factor model. *Environmental Science and Technology.* **35**(9): 1709-1715.

Wu, C.X., A.L. Spongberg, and J.D. Witter (2008). Determination of the persistence of pharmaceuticals in biosolids using liquid-chromatography tandem mass spectrometry. *Chemosphere.* **73**(4): 511-518.

Wu, C.X., A.L. Spongberg, and J.D. Witter (2009). Adsorption and Degradation of Triclosan and Triclocarban in Soils and Biosolids-Amended Soils. *Journal of Agricultural and Food Chemistry.* **57**(11): 4900-4905.

Xia, K., A. Bhandari, K. Das, and G. Pillar (2005). Occurrence and fate of pharmaceuticals and personal care products (PPCPs) in biosolids. *Journal of Environmental Quality.* **34**(1): 91-104.

Xia, K., L.S. Hundal, K. Kumar, K. Armbrust, A.E. Cox, and T.C. Granato (2010). Triclocarban, triclosan, polybrominated diphenyl ethers, and 4-nonylphenol in biosolids and soil receiving 33-year biosolids application. *Environmental Toxicology and Chemistry.* **29**(3): 597-605.

Xu, S.H., R.G. Lehmann, J.R. Miller, and G. Chandra (1998). Degradation of polydimethylsiloxanes (silicones) as influenced by clay minerals. *Environmental Science and Technology.* **32**(9): 1199-1206.

Xu, W.H., G. Zhang, X.D. Li, S.C. Zou, P. Li, Z.H. Hu, and J. Li (2007). Occurrence and elimination of antibiotics at four sewage treatment plants in the Pearl River Delta (PRD), South China. *Water Research.* **41**(19): 4526-4534.

Yamada, T., Y. Takahama, and Y. Yamada (2009). Isolation of Pseudomonas sp. strain HB01 which degrades the persistent brominated flame retardant gamma-hexabromocyclododecane. *Bioscience, Biotechnology and Biochemistry.* **73**(7): 1674-8.

Yamamoto, H., H.M. Liljestrand, Y. Shimizu, and M. Morita (2003). Effects of physical-chemical characteristics on the sorption of selected endocrine disruptors by dissolved organic matter surrogates. *Environmental Science and Technology.* **37**(12): 2646-2657.

Yang, J.J. and C.D. Metcalfe (2006). Fate of synthetic musks in a domestic wastewater treatment plant and in an agricultural field amended with biosolids. *Science of the Total Environment.* **363**(1-3): 149-165.

Ying, G.G. and R.S. Kookana (2005). Sorption and degradation of estrogen-like-endocrine disrupting chemicals in soil. *Environmental Toxicology and Chemistry.* **24**(10): 2640-2645.

Ying, G.G. and R.S. Kookana (2007). Triclosan in wastewaters and biosolids from Australian wastewater treatment plants. *Environment International.* **33**(2): 199-205.

Ying, G.G., R.S. Kookana, and P. Dillon (2003). Sorption and degradation of selected five endocrine disrupting chemicals in aquifer material. *Water Research.* **37**(15): 3785-3791.

Ying, G.G., R.S. Kookana, and Y.J. Ru (2002). Occurrence and fate of hormone steroids in the environment. *Environment International.* **28**(6): 545-551.

Ying, G.G., X.Y. Yu, and R.S. Kookana (2007). Biological degradation of triclocarban and triclosan in a soil under aerobic and anaerobic conditions and comparison with environmental fate modelling. *Environmental Pollution.* **150**(3): 300-305.

Yoo, H., J.W. Washington, T.M. Jenkins, and E.L. Libelo (2009). Analysis of perfluorinated chemicals in sludge: Method development and initial results. *Journal of Chromatography A.* **1216**(45): 7831-7839.

Yoshimoto, T., F. Nagai, J. Fujimoto, K. Watanabe, H. Mizukoshi, T. Makino, K. Kimura, H. Saino, H. Sawada, and H. Omura (2004). Degradation of Estrogens by Rhodococcus zopfii and Rhodococcus equi Isolates from Activated Sludge in Wastewater Treatment Plants. *Applied and Environmental Microbiology.* **70**(9): 5283-9.

Yu, Z.Q. and W.L. Huang (2005). Competitive sorption between 17 alpha-ethinyl estradiol and naphthalene/phenanthrene by sediments. *Environmental Science and Technology.* **39**(13): 4878-4885.

Yu, Z.Q., B.H. Xiao, W.L. Huang, and P. Peng (2004). Sorption of steroid estrogens to soils and sediments. *Environmental Toxicology and Chemistry.* **23**(3): 531-539.

Zerzghi, H., C. Gerba, J. Brooks, and I. Pepper (2010). Long-term effects of land application of class B biosolids on the soil microbial populations, pathogens, and activity. *J Environ Qual.* **39**(1): 402-8.

Zhang, H. and C.H. Huang (2007). Adsorption and oxidation of fluoroquinolone antibacterial agents and structurally related amines with goethite. *Chemosphere.* **66**(8): 1502-1512.

Zhou, J.L. (2006). Sorption and remobilization behavior of 4-tert-octylphenol in aquatic systems. *Environmental Science and Technology.* **40**(7): 2225-2234.

Zurhelle, G., M. Petz, E. Mueller-Seitz, and E. Siewert (2000). Metabolites of oxytetracycline, tetracycline, and chlortetracycline and their distribution in egg white, egg yolk, and hen plasma. *Journal of Agricultural and Food Chemistry.* **48**(12): 6392-6396.

WASTEWATER UTILITY

Alabama
Montgomery Water Works & Sanitary Sewer Board

Alaska
Anchorage Water & Wastewater Utility

Arizona
Avondale, City of
Glendale, City of, Utilities Department
Mesa, City of
Peoria, City of
Phoenix Water Services Dept.
Pima County Wastewater Management
Safford, City of
Tempe, City of

Arkansas
Little Rock Wastewater Utility

California
Central Contra Costa Sanitary District
Corona, City of
Crestline Sanitation District
Delta Diablo Sanitation District
Dublin San Ramon Services District
East Bay Dischargers Authority
East Bay Municipal Utility District
El Dorado Irrigation District
Fairfield-Suisun Sewer District
Fresno Department of Public Utilities
Inland Empire Utilities Agency
Irvine Ranch Water District
Las Gallinas Valley Sanitary District
Las Virgenes Municipal Water District
Livermore, City of
Los Angeles, City of
Los Angeles County, Sanitation Districts of
Napa Sanitation District
Novato Sanitary District
Orange County Sanitation District
Palo Alto, City of
Riverside, City of
Sacramento Regional County Sanitation District
San Diego Metropolitan Wastewater Department, City of
San Francisco, City & County of
San Jose, City of
Santa Barbara, City of
Santa Cruz, City of
Santa Rosa, City of
South Bayside System Authority
South Coast Water District
South Orange County Wastewater Authority
South Tahoe Public Utility District
Stege Sanitary District
Sunnyvale, City of
Union Sanitary District
West Valley Sanitation District

Colorado
Aurora, City of
Boulder, City of
Greeley, City of
Littleton/Englewood Water Pollution Control Plant
Metro Wastewater Reclamation District, Denver

Connecticut
Greater New Haven WPCA
Stamford, City of

District of Columbia
District of Columbia Water & Sewer Authority

Florida
Broward, County of
Fort Lauderdale, City of
Jacksonville Electric Authority (JEA)
Miami-Dade Water & Sewer Authority
Orange County Utilities Department
Pinellas, County of
Reedy Creek Improvement District
Seminole County Environmental Services
St. Petersburg, City of
Tallahassee, City of
Toho Water Authority
West Palm Beach, City of

Georgia
Atlanta Department of Watershed Management
Augusta, City of
Clayton County Water Authority
Cobb County Water System
Columbus Water Works
Fulton County
Gwinnett County Department of Public Utilities
Savannah, City of

Hawaii
Honolulu, City & County of

Idaho
Boise, City of

Illinois
Decatur, Sanitary District of
Greater Peoria Sanitary District
Kankakee River Metropolitan Agency
Metropolitan Water Reclamation District of Greater Chicago
Wheaton Sanitary District

Indiana
Jeffersonville, City of

Iowa
Ames, City of
Cedar Rapids Wastewater Facility
Des Moines, City of
Iowa City

Kansas
Johnson County Wastewater
Unified Government of Wyandotte County/ Kansas City, City of

Kentucky
Louisville & Jefferson County Metropolitan Sewer District
Sanitation District No. 1

Louisiana
Sewerage & Water Board of New Orleans

Maine
Bangor, City of
Portland Water District

Maryland
Anne Arundel County Bureau of Utility Operations
Howard County Bureau of Utilities
Washington Suburban Sanitary Commission

Massachusetts
Boston Water & Sewer Commission
Massachusetts Water Resources Authority (MWRA)
Upper Blackstone Water Pollution Abatement District

Michigan
Ann Arbor, City of
Detroit, City of
Holland Board of Public Works
Saginaw, City of
Wayne County Department of Environment
Wyoming, City of

Minnesota
Rochester, City of
Western Lake Superior Sanitary District

Missouri
Independence, City of
Kansas City Missouri Water Services Department
Little Blue Valley Sewer District
Metropolitan St. Louis Sewer District

Nebraska
Lincoln Wastewater & Solid Waste System

Nevada
Henderson, City of
Las Vegas, City of
Reno, City of

New Jersey
Bergen County Utilities Authority
Ocean County Utilities Authority

New York
New York City Department of Environmental Protection

North Carolina
Charlotte/Mecklenburg Utilities
Durham, City of
Metropolitan Sewerage District of Buncombe County
Orange Water & Sewer Authority
University of North Carolina, Chapel Hill

Ohio
Akron, City of
Butler County Department of Environmental Services
Columbus, City of
Metropolitan Sewer District of Greater Cincinnati
Montgomery, County of
Northeast Ohio Regional Sewer District
Summit, County of

Oklahoma
Oklahoma City Water & Wastewater Utility Department
Tulsa, City of

Oregon
Albany, City of
Clean Water Services
Eugene, City of
Gresham, City of
Portland, City of Bureau of Environmental Services
Lake Oswego, City of
Oak Lodge Sanitary District
Water Environment Services

Pennsylvania
Hemlock Municipal Sewer Cooperative (HMSC)
Philadelphia, City of
University Area Joint Authority

South Carolina
Charleston Water System
Mount Pleasant Waterworks & Sewer Commission
Spartanburg Water

Tennessee
Cleveland Utilities
Murfreesboro Water & Sewer Department
Nashville Metro Water Services

Texas
Austin, City of
Dallas Water Utilities
Denton, City of
El Paso Water Utilities

Fort Worth, City of
Houston, City of
San Antonio Water System
Trinity River Authority

Utah

Salt Lake City Corporation

Virginia

Alexandria Sanitation Authority
Arlington, County of
Fairfax, County of
Hampton Roads Sanitation District
Hanover, County of
Henrico, County of
Hopewell Regional Wastewater Treatment Facility
Loudoun Water
Lynchburg Regional Wastewater Treatment Plant
Prince William County Service Authority
Richmond, City of
Rivanna Water & Sewer Authority

Washington

Everett, City of
King County Department of Natural Resources
Seattle Public Utilities
Sunnyside, Port of
Yakima, City of

Wisconsin

Green Bay Metro Sewerage District
Kenosha Water Utility
Madison Metropolitan Sewerage District
Milwaukee Metropolitan Sewerage District
Racine, City of
Sheboygan Regional Wastewater Treatment
Wausau Water Works

Water Services Association of Australia

ACTEW Corporation
Barwon Water
Central Highlands Water
City West Water
Coliban Water Corporation
Cradle Mountain Water
Gippsland Water
Gladstone Area Water Board
Gold Coast Water
Gosford City Council
Hunter Water Corporation
Logan Water
Melbourne Water
Moreton Bay Water
Onstream
Power & Water Corporation
Queensland Urban Utilities
SEQ Water
South Australia Water Corporation
Sunshine Coast Water
Sydney Catchment Authority
Sydney Water
Unity Water
Wannon Regional Water Corporation
Watercare Services Limited (NZ)
Water Corporation
Western Water
Yarra Valley Water

Canada

Edmonton, City of/Edmonton Waste Management Centre of Excellence
Lethbridge, City of
Regina, City of, Saskatchewan
Toronto, City of, Ontario
Winnipeg, City of, Manitoba

STORMWATER UTILITY

California

Fresno Metropolitan Flood Control District
Los Angeles, City of, Department of Public Works
Monterey, City of
San Francisco, City & County of
Santa Rosa, City of
Sunnyvale, City of

Colorado

Aurora, City of
Boulder, City of

Florida

Orlando, City of

Iowa

Cedar Rapids Wastewater Facility
Des Moines, City of

Kansas

Lenexa, City of
Overland Park, City of

Kentucky

Louisville & Jefferson County Metropolitan Sewer District

Maine

Portland Water District

North Carolina

Charlotte, City of, Stormwater Services

Pennsylvania

Philadelphia, City of

Tennessee

Chattanooga Stormwater Management

Texas

Harris County Flood Control District, Texas

Washington

Bellevue Utilities Department
Seattle Public Utilities

STATE

Connecticut Department of Environmental Protection
Kansas Department of Health & Environment
New England Interstate Water Pollution Control Commission (NEIWPCC)
Ohio Environmental Protection Agency
Ohio River Valley Sanitation Commission
Urban Drainage & Flood Control District, CO

CORPORATE

ADS LLC
Advanced Data Mining International
AECOM
Alan Plummer & Associates
Alpine Technology Inc.
Aqua-Aerobic Systems Inc.
Aquateam–Norwegian Water Technology Centre A/S
ARCADIS
Associated Engineering
Bernardin Lochmueller & Associates
Black & Veatch
Blue Water Technologies, Inc.
Brown & Caldwell
Burgess & Niple, Ltd.
Burns & McDonnell
CABE Associates Inc.
The Cadmus Group
Camp Dresser & McKee Inc.
Carollo Engineers Inc.
Carpenter Environmental Associates Inc.
CET Engineering Services
CH2M HILL
CRA Infrastructure & Engineering
CONTECH Stormwater Solutions
D&B/Guarino Engineers, LLC
Damon S. Williams Associates, LLC
Ecovation
EMA Inc.
Environmental Operating Solutions, Inc.
Environ International Corporation
Fay, Spofford, & Thorndike Inc.
Freese & Nichols, Inc.
ftn Associates Inc.
Gannett Fleming Inc.
Garden & Associates, Ltd.
Geosyntec Consultants
GHD Inc.
Global Water Associates
Greeley and Hansen LLC
Hazen & Sawyer, P.C.
HDR Engineering Inc.
HNTB Corporation
Hydromantis Inc.
HydroQual Inc.
Infilco Degremont Inc.
Jason Consultants LLC Inc.
Jordan, Jones, & Goulding Inc.
KCI Technologies Inc.
Kelly & Weaver, P.C.
Kennedy/Jenks Consultants
Larry Walker Associates
LimnoTech Inc.
Lombardo Associates, Inc.
The Low Impact Development Center Inc.
Malcolm Pirnie Inc.
Material Matters, Inc.
McKim & Creed
MWH
NTL Alaska, Inc.
O'Brien & Gere Engineers Inc.
Odor & Corrosion Technology Consultants Inc.
Parametrix Inc.
Parsons
Post, Buckley, Schuh & Jernigan
Praxair, Inc.
RMC Water & Environment
Ross & Associates Ltd.
SAIC
Siemens Water Technologies
The Soap & Detergent Association
Smith & Loveless, Inc.
Southeast Environmental Engineering, LLC
Stone Environmental Inc.
Stratus Consulting Inc.
Synagro Technologies Inc.
Tetra Tech Inc.
Trojan Technologies Inc.
Trussell Technologies, Inc.
URS Corporation
Wallingford Software
Westin Engineering Inc.
Wright Water Engineers
Zoeller Pump Company

INDUSTRY

American Electric Power
American Water
Anglian Water Services, Ltd.
Chevron Energy Technology
The Coca-Cola Company
Dow Chemical Company
DuPont Company
Eastman Chemical Company
Eli Lilly & Company
InsinkErator
Johnson & Johnson
Merck & Company Inc.
Procter & Gamble Company
Suez Environnment
United Utilities North West (UUNW)
United Water Services LLC
Veolia Water North America

Note: List as of 4/20/10

IWA Publishing's authorised EU representative for General Product Safety Regulations is Diane D'Arras, 15 rue Duret, 75116 Paris, France, e-mail: safety@iwap.co.uk.

Printed and bound by CPI Group (UK) Ltd, Croydon, CR0 4YY
17/09/2026
02218804-0002